D0847840

Nature Underfoot

Nature Underfoot

Living with Beetles, Crabgrass, Fruit
Flies, and Other Tiny Life Around Us

John Hainze

Illustrated by Angela Mele

Yale UNIVERSITY PRESS

New Haven & London

Yale University Press books may be purchased in quantity for educational,
business, or promotional use. For information, please e-mail
sales.press@yale.edu (U.S. office) or sales@yaleup.co.uk (U.K. office).

Set in Janson Roman type by Integrated Publishing Solutions,
Grand Rapids, Michigan.
Printed in the United States of America.

ISBN 978-0-300-24278-2 (hardcover : alk. paper)
Library of Congress Control Number: 2019941039
A catalogue record for this book is available from the British Library.

This paper meets the requirements of ANSI/NISO Z39.48-1992
(Permanence of Paper).

10 9 8 7 6 5 4 3 2 1

To my parents, Bobbie and Eugene Hainze

I believe a leaf of grass is no less than the journey-work of the stars . . .

—WALT WHITMAN (1819–1892)

Contents

Contents

Preface

The Food and Agriculture Organization's estimate of pesticide use in 2016 found that nearly nine billion pounds (roughly four million kilograms) of pesticides were applied around the world that year. This is a truly gargantuan total of sophisticated chemicals that kill their targets, principally weeds, insects, and fungi, in micro amounts. There is also collateral damage, as other innocent-bystander organisms are destroyed in the process. Perhaps hundreds of trillions of individual weeds and insects were killed that year in what is a continuous and ongoing assault on the biosphere. Of course, much of the use of pesticides is agricultural, where these chemicals are spread over large areas, resulting in broad swaths of generally indiscriminate insect and plant destruction. But pesticides are also widely used in and around human homes and buildings. In a report issued in 2017, the United States Environmental Protection Agency (EPA) estimated that, in 2012, eighty-eight million households in the United States applied pesticides of one form or another. This figure represents more than 70 percent of United States households using chemical pesticides. Some may say that the broad use of pesticides is necessary to protect our food and fiber, to prevent disease, and to maintain our homes. This book is intended to challenge the ease with which we turn to killing the organisms that many refer to as pests.

A truly astounding amount of killing is taking place in growers' fields, in towns, and in cities, and nearly all of us take part in the massacre. Insect and plant lives were cut short in many more than 70 percent of United States households—probably 100 percent, since a blunt instrument, such as a rolled-up newspaper or fly swatter, is often used on insects, and plants may be pulled from the ground or chopped down. Insecticide and herbicide products on store shelves provide long lists of arthropods (insects, spiders, millipedes, and the like) and weeds that they kill—many of which pose little threat to humans. The extensive use of chemical pesticides and other methods of extermination in United States households reflects an animosity toward smaller living things that is often based on a misunderstanding of their effects in homes and gardens. The heedless destruction of their habitat through expanding human land use further suggests that these organisms are generally held in low regard. Our relationship with them is a growing concern in the context of the wide-ranging impact of humans on the earth today, a period that has come to be called the Anthropocene epoch, to reflect the recent fundamental changes to the planet wrought by human activity.

It is a time when insects are in spectacular decline around the world, which the popular press describes as the insect apocalypse. In Germany, citizen scientists documented a loss of as much as 75 percent of the mass of flying insects over a twenty-seven-year period. They recorded these losses even in areas protected from agriculture and development. In the Puerto Rico rainforest, researchers determined that arthropods had declined by up to sixty times between 1976 and 2012. No one is absolutely certain why the flying insects are declining in Germany, but pesticides are the suspected cause. The authors of the Puerto Rico study believe that climate change is to blame for the loss of arthropods there. A review article in 2019 concluded that globally more than 40 percent of insect species are threatened with extinction. Bees, butterflies, dragonflies, and dung beetles are mentioned as taxa that are experiencing very significant decline—all of which are represented in this book. The authors of the review say that habitat loss to intensive agriculture and urbanization is the biggest factor in the decrease in insects around the world. Other drivers of global insect losses include pollution from fertilizers and pesticides, diseases, introduction of nonnative species, and climate change. Either directly or inadvertently, human beings are killing a lot of plants and arthropods, to the point of extinguishing many species and impoverishing the earth's biodiversity. Clearly, it is time to reckon with how we relate to these smaller living beings. Something has to change, and I think this begins with our attitudes toward nature writ small.

I believe the current crisis faced by these organisms that are so critical to the planet's ecosystems calls for reflection on how small plants and animals affect us, and we affect them. In writing this book, I hope to raise awareness and increase our appreciation for these living beings. It is an effort to raise their profile by relating features of their fascinating existence—especially those of plants and animals that live in close proximity to us in urban and suburban settings. For I think we must first confront our attitudes toward them where they are most frequently encountered—in the environments that we have constructed. It seems high time to consider our responsibilities toward our small neighbors. Although it may seem counterintuitive, I submit that humans ultimately have a moral obligation toward even these often-unloved or ignored organisms.

I approach the material through my training as an entomologist, from experience as an insider working in the pesticide industry, and from a perspective gleaned from religion and philosophy. Even entomologists, who have developed a healthy appreciation or even admiration for insects, may exhibit a disregard for the lives of their research subjects. I must admit that it did not cross my mind to ask whether my research goals were worthy of the insect lives I sacrificed. I have since become more conscious of the unnecessary loss of the plant and arthropod life that surrounds us.

The book is intended to be accessible to anyone with an interest in nature. I hope it can contribute to rethinking our relationship with the smaller life around us. In that respect, the book is meant to be scientifically accurate but also inviting. Scientific names and terms are used in addition to occasional anthropomorphic descriptions of creatures or their behaviors. I have chosen to write about aspects of the lives of plants, insects, and their kin that offer examples of important relationships of humans with these beings. The material is taken selectively from the scientific literature. It contains much natural history, but it is not solely about natural history. In this, the book differs from other similar literature. Instead, it uses natural history to broaden the awareness of smaller plants and animals, and to support an argument about the value of and our responsibilities toward underappreciated life. The aim is to change our thinking about less charismatic nature: the life we find underfoot in our everyday lives.

Acknowledgments

This book had its beginnings in the environmental writing seminar Fred Strebeigh teaches at the Yale University School of Forestry and Environmental Sciences. The time spent with him and the students in that class was invaluable, and his generous guidance helped immeasurably.

I am grateful to those who read portions of the book and provided their insight, including Walt Goodman, Fred Simmons, Mike Pellegrino, and six anonymous reviewers. Additionally, I owe much in this book to discussions in a reading course on environmental ethics with Fred Simmons and Mike Pellegrino. Merrilee MacLean provided the sharp eyes of a legal scholar in proofreading the text.

Mary Evelyn Tucker and John Grim and their efforts in the Forum on Religion and Ecology at Yale University have been an ongoing inspiration. Their kind support and encouragement were fundamental to my attempt at this project.

I feel very fortunate to have had the opportunity to work with Jean Thomson Black at Yale University Press. She understood my vision, and her perceptive direction resulted in its publication. Michael Deneen at the press ably contributed to moving the project forward. Phillip King's editing provided clarity and made the book infinitely more readable.

A special thanks to Angela Mele, whose graceful and accurate illustrations lead each chapter. She is a pleasure to work with and was a discerning and reliable partner for the project.

The story of the political machinations surrounding the American burying beetle in Chapter 5 first appeared in an article in *The Progressive* online, "Gaming the Endangered Species Act: The Case of the American Burying Beetle" (2017; https://progressive .org/dispatches/gaming-the-endangered-species-act-the-case -of-the-american-b/).

Support for this project was provided through the Issachar Fund Writer's Retreat, allowing a period of focused writing in which the core of the book was completed.

I express heartfelt thanks and love to my family, Jane, Anna, Emily, Sasha, and Erik, and to my sister, Jennifer, for their confidence and encouragement. They also read and provided helpful commentary on the text.

Finally, my gratitude and love for her unstinting patience and innumerable contributions to this book to my wife, Joan.

Nature Underfoot

Introduction

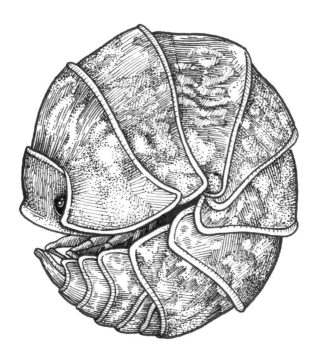

I ASSEMBLED A MODEL of the creature from the black lagoon when I was eight or nine years old. It loomed large in my imagination and was the stuff of nightmares—human-like, standing on two legs, but with a green scaly body, webbed feet and hands, and fringed gills emerging from its neck. Non-human nature, like this scaly creature emerging from the depths, is often represented in the popular imagination as something to be feared. A more recent film, *The Shape of Water*, introduces an amphibious creature that is based on my model from the earlier film, *The Creature from the Black Lagoon*.[1] This creature also seems frightening and is a fierce carnivore, devouring a pet cat at one point in the movie. But there is something more to this creature than its ferocity: a love story develops between the creature and a woman who cleans the facility where it is imprisoned. How could the woman have feelings for something so very different in appearance and habit from humans? Despite this strangeness, as the film progresses, we also begin to care for the creature, and it seems outrageous when the creature's captor tortures it with an electric shock stick. We feel that the creature, regardless of its appearance and unusual behavior, should be treated more humanely. Can these human feelings for distinctly non-human life extend to the real world?

Happily, we don't have to look far for armored creatures with a gill-like breathing apparatus. We can find them roaming our homes. They are pillbugs and their relatives, the sowbugs. They get around on seven pairs of legs and are small—maybe half an inch in length. So a love affair of the sort depicted in *The Shape of Water* is not likely with a pillbug. But, as I will suggest, affection for these creatures and other tiny organisms that intersect our lives is definitely possible. Scientists refer to animals like pillbugs and sowbugs as isopods—a kind of crustacean, the group of animals that also contains shrimp and crabs. The isopods originally lived only in water, but more than fifty million years ago some began to move to dry land. The gill-like breathing organ is a vestige of their aquatic origins. Unsurprisingly, given their ancestry, pillbugs require moist conditions to survive, and they are often found under stones, logs, and leaf litter. Our basements may provide sufficient humidity for pillbugs when they venture indoors. Most other parts of our homes, though, are generally too dry for pillbugs to survive. As a result, many of us have witnessed the sad scene in which dead, curled-up pillbugs litter the floor. People often react to the presence of pillbugs by using a vacuum to remove them, whether dead or alive. Yet gently returning live pillbugs to a suitably moist place outdoors is doing the environment a favor. They are great recyclers and important decomposers of dead leaves. They help ensure that we are not knee deep in leaf litter and that the soil is replenished with nutrients. There is definitely something to love about these armored creatures whose ancestors emerged from a black lagoon fifty million years ago to take up life on land.

I remember having an early fascination with pillbugs when I was young. Some people call them potato bugs, and others call them roly-polies. It's not unusual to have multiple common names for the same organism or group of organisms. If any of those names spark recognition, then you'll remember that these creatures roll into a ball when threatened—thus the name roly-poly. They also do this to preserve moisture—that is why many are found in a curled-up position when they die as a result of dehydration. I found prodding pillbugs to make them roll into a ball a captivating activity when I was young. Sometimes they didn't cooperate, which was puzzling. I now know that the similar-looking sowbugs are physically unable to roll into a ball. Instead, they play dead when disturbed—a behavior that is much less appealing. I'm afraid that I may have tortured many sow-bugs, expecting them to roll up—not appreciating that they were different animals from pillbugs. Both pillbugs and sowbugs frequently inhabit our homes and yards, and so are forms of nature we can readily enjoy—but they're probably better off if you avoid prodding them.

In discussing how we perceive nature and ecology, Aldo Leopold wrote, "The weeds in a city lot convey the same lesson as the redwood."[2] I understand him to mean that weeds and redwoods both find ways to flourish and reproduce and that we can learn much from each about how they interact with their environment and other organisms. Each plant has evolved ways of dealing with competitors, and acquiring the nutrients, water, and sunlight they need—their strategies differ but the basic problem of how to grow and reproduce is the same. Weeds may grow and reproduce quickly in disturbed, transitional environments. Redwoods come to tower over and dominate a more stable assemblage of plants during a lifetime that far exceeds that of human beings (up to two thousand years). Thinking solely about their ecology, the scrawny weeds in the city lot can be every bit as interesting as the grand redwood and an important part of nature. Nature is often thought of as life that occurs beyond human habitation, and if there is nature in cities, then it is found in parks and gardens. Yet we are animals too, and thus part of nature. And nature, like the weeds in a city lot, is present throughout human-built environments. This is the kind of nature that any of us might encounter on a daily basis. Thinking about it in this way, we are not separated from nature; it surrounds us. We just may not be aware of it.

Awareness is fundamental to a new way of communing with nature that originated in Japan. It's called forest bathing. As I understand the practice, it is the process of slowly walking in the forest, being fully present to all the sights, smells, and sounds there. I think it's a wonderful idea, and based on my own experience can relate to the feeling of being at peace in a forested setting. I wonder, though, if we could have a similar experience in our own neighborhood. Could we be present to the sounds of crickets and bumble bees; to the earthy smell of a vacant lot; to the sight of bright yellow dandelions in the lawn or ants busying themselves on the sidewalk? Maybe human-made sounds would be too intrusive, but I also think there is much that we miss in our local environment.

As an entomologist, I chose to make a career of learning about insects. But I also used that knowledge against my chosen subjects. I worked for many years on products such as Off! mosquito repellent and Raid insect killer. I believed that these products improved human lives and was involved with other similarly dedicated scientists in making the products safer and more effective. Mosquito repellents can be an important tool in preventing the transmission of certain mosquito-borne diseases, like West Nile virus. Some of the insect killer products could be used to reduce the number of cockroaches in homes—and high populations of cockroaches in a home are known to be a factor that may contribute to childhood asthma. But these tools may be misused, too. Some people use insecticides to kill pillbugs and sowbugs—animals that have difficulty surviving in a home anyway and that really cause no harm to humans. Of course, people are also using these products without training, and may apply them in ways that are unsafe for their families or for the environment. And, I ask myself, why do we put fragrance in our insecticides? I'm not so sure that using insecticides should be a pleasing sensory experience. In fact, some countries don't allow it. Insecticide products should be a tool of last resort—used only when human well-being is truly threatened.

Not too long ago, I had the chance to spend a couple of years studying religion and ethics at Yale University—quite a change from working in the biological sciences. It was an opportunity to think about how human beings relate to each other and how we relate to the environment. This book is an outgrowth of that time and conversations at school. I am most interested in how we think about the responsibilities we have to the environment. Given my background, I gravitate to understanding how humans relate to what many might think of as lesser inhabitants of the environment. Since I had been working on insects that affect humans, I did a lot of thinking about the living things that we find in our human-made environment. These organisms often escape our notice, or if we do notice them it's when we encounter them inside our home or in our garden. My aim is to raise awareness of these smaller living things in our built environment and beyond, and to develop a more widespread appreciation for them. Through appreciation I hope we will begin to value these creatures, and consider more thoughtfully the moral obligations we may have toward them.

There is a lot of discussion these days about the present geologic epoch. Many agree that we are in a new epoch—one that reflects the impact of human beings on the earth. Scientists call it the Anthropocene epoch. There is disagreement as to exactly when it started. Some say it started about ten thousand years ago when humans began to practice agriculture. Others say it will be marked by the first detonation of the atomic bomb, and the radioactive materials left behind. Still others say it will be delineated by the presence of plastics in the geological strata. The idea is that future geologists will encounter these markers as they look at the earth's layers, just as present-day geologists identify past epochs by the presence of certain fossils, rocks, or chemicals in a layer. The Anthropocene is thus recognized as an epoch in which the changes produced by human beings create a greater impact on the earth than those occurring naturally.

The environments most significantly affected by humans are our cities and towns, which are full of structures, imported nonnative plants, and domestic animals. I often wonder how these environments look to plants and animals that encounter them. Ecologists characterize urban environments as patchy. Much of the surface of urban areas, the streets, parking lots, sidewalks, driveways, and roofs, are paved and impervious—places where water can't penetrate, and plants can't grow. But there are patches of open soil in parks, yards, and vacant lots that do support the growth of plants. A variety of microscopic to small animals also make their homes in these vegetated islands. Seeds may be blown or dropped into cracks and crevices of sidewalks, streets, and parking lots, and plants will spring up in the most unlikely locations. Much of the urban outdoor environment is maintained in an open condition like lawns or vacant lots, where plants that grow rapidly and enjoy full sun—pioneer plants like dandelions and crabgrass—are successful. The exterior of buildings must appear to plants and animals as vertical cliff faces. Organisms may find purchase in indentations or gaps in the upright surfaces. Mold or moss may grow on the surface itself when conditions are right. I imagine the interior of buildings must seem somewhat cave-like to other organisms—though apart from basements, building interiors are generally a very dry environment. Insects, centipedes, millipedes, pillbugs, and many others find their way into our homes through open doors and windows, and structural cracks and crevices. Sustenance is available there in the form of stored goods, other small organisms, and food spills. What is food for them may be surprising, including the glue on envelopes or even the paper envelopes themselves for certain insects. But the dryness of the environment makes it hard for these animals to survive. Pillbugs are one such example, requiring a moist environment for survival—usually in the basement or bathroom. Organisms that can manage the dry conditions live in walls, under furniture, in storage areas, under sinks, and any other area away from human traffic. Many find a microenvironment that provides enough moisture to survive—like dust mites, which find areas of higher humidity lurking in the forest of fibers at the base of carpets or on upholstered furniture.

How have human relationships with these small living things changed in the Anthropocene? You may not feel that you have a relationship at all with such tiny and insignificant creatures, but you do. In terms of proximity, many of the small organisms that we will encounter in this book are our closest neighbors. They are a kind of domestic wildlife in and around our homes. No matter how hard we work to produce a pristine environment, nature intrudes. Many of the plants and animals that live close to humans in the Anthropocene have also traveled around the world with us. The rapidity and frequency of the introduction of species to new locales is a peculiar feature of the Anthropocene. It is another way in which we transform the earth—sometimes purposefully, but more often accidentally. Depending on how it's defined, the Anthropocene may also correlate with what is being called the sixth great extinction. It is generally agreed that the present high rate of extinctions is the result of human activity. We shape the earth with each small loss to its ecosystems, evidenced by a missing decomposer, pollinator, or plant that held the soil in place.

We are accompanied in the Anthropocene by a group of living things that flourish even in the environments that we create for ourselves. I hope that learning more about them will encourage a shift in how we relate to these organisms. Maybe we will find that nature bathing is achievable at home. Maybe we will also find it possible to share space with the more innocuous members of this group of small neighbors. These plants and animals, innocuous or not, bring a richness to our daily lives that is readily accessible in our living rooms and lawns. It is nature underfoot, where fruit flies hover in the kitchen, crabgrass and dandelions decorate the lawn, and silverfish may be observed meeting shyly in the basement.

ONE

Anthropocene Winners

THEY MEET IN THE BASEMENT where it is suitably damp and dark, behind boxes of discarded clothing and toys. The first touch is electric. It arises from the mutual contact of antennae. Don't worry. It's consensual. They touch antennae, retreat and return, and retreat and return again. They may bob their heads. The feeling of mutual attraction grows and they go for a stroll where the damp cement floor meets cinderblock wall. Perhaps a chase is involved, male chasing female or vice versa. As they become more confident in their relationship, he lays a mat of silken threads before her where he provides a nuptial gift. When she reaches the mat the threads stimulate her to search for the gift with her tail and, when found, she grasps the package of sperm with the ovipositor at the tip of her tail. On the whole, it's a rather chaste interaction of male and female that lasts about half an hour. Sperm are transferred without intercourse. So goes the mating ritual of silverfish (family Lepismatidae), a perfectly ordinary household insect that is found in the largest cities and down on the farm. One can find silverfish in locations all over the house in habitats that we create for them. We know them by their slippery quick movements, their elongate, silvery appearance, long antennae, and three "tails."

I used to be a killer of insects such as these silverfish. As an entomologist, I have plenty of hemolymph (insect blood) on my hands. My colleagues and I cooked up recipes and new ways of dispatching crawling, flying, oozing, and rooted life forms. We made Raid bug killers—the tool of choice for the extermination of all manner of living things around the world. We experimented with a variety of -cides, including insecticides, acaricides, molluscicides, rodenticides, herbicides, fungicides, and bactericides. Our efforts were prodigious and unrelenting, yet few of the organisms that users target with these products represent much of a threat to humans. The rest are an inconvenience or objects of a puzzling fear and loathing disproportionate to their effect. Sometimes spilling blood of the insect variety involved losing a little of my own. The insects and I both took some losses when I worked on methods to prevent mosquito bites. Sadly, I took even more insect lives in finding ways to limit cockroach and ant invasions in human homes. But should human needs always take precedence?

I faced this quandary recently when I pulled a cookbook from the shelf and dislodged a silverfish from its hiding place. For a split second—silverfish are fast—the insect's fate was suspended as I considered the inconvenience of a silverfish roaming through my breakfast cereal and my desire to respect a fellow traveler on the planet. I brought my hand down with a whack onto the countertop. I'm still a killer. I can wax eloquent about the virtues of silverfish, but in that moment of truth I'm afraid I acted solely in the interest of my cereal. In the days that followed, I vacillated between regret for the killing and resignation to my act. At least I dispatched the silverfish with my bare hand. No poison was involved, no impersonal newspaper. The blow was human skin on silverfish scales. It certainly would have been more respectful had

I simply let it go on its way. But having missed that opportunity, perhaps another form of respect, though of little consolation to my silverfish, is to understand these creatures a little bit better.

Silverfish are venerable animals. Picture a humid and swampy landscape in the Carboniferous period, over 300 million years ago. Silverfish would have been foraging in the leaf litter among ferns and giant horsetail plants, ancient trees towering up to one hundred feet above them. An amphibian, which began to appear on land at this time, might have chased the silverfish around roots and under fallen leaves looking for a meal. The silverfish line is one of the oldest among the insects. For perspective, dinosaurs began to appear about 200 million years ago, and the human species evolved about 200,000 years ago. Surprisingly, silverfish today are very similar to their earliest ancestors. Their lives did not require wings, and they remain a wingless group today. Not many organisms have withstood the vicissitudes of natural selection relatively unchanged for hundreds of millions of years. It seems that they were structured for success in the evolutionary arms race from their very beginnings. They survived three mass extinctions, including the most recent, which wiped out the dinosaurs.

Why are silverfish so successful? Silverfish have small, flattened bodies that enable them to hide in cracks and crevices, beneath bark or under stones, avoiding predators and conserving internal moisture. The ability to use a variety of food sources, including decaying plant material and even their own molted skins, certainly helps. Behavioral flexibility is another likely success factor. Scientists have found that silverfish are able to learn a complex maze, though they don't retain the knowledge for long. Thus the ability to learn safe foraging paths might also contribute to their survival. Most likely, evolutionary success is a complex combination of these and other features of these remarkable insects. I admire their longevity and their triumph over evolutionary forces. Recently, very recently in evolutionary terms, they have become fellow travelers with humans and are closely associated with human habitation. They are among the winners of the Anthropocene, the current epoch in which human activities, rather than nature, are the main factor in changing and shaping the earth. The global success of silverfish today is tightly intertwined with the environments that humans unsuspectingly create for them, often in what would otherwise be forbidding locations. We build for them and transport them when we move books, boxes, and furniture. They are a little bit of nature that we live with every day. We meet them as successful life forms but may find them to be an annoyance. They certainly can be. Knowing a little more about silverfish, though, may tell us something more about ourselves and the world we've created, a world that benefits an assortment of living beings whose survival strategies match the environments we construct.

Our perceptions of animals like silverfish are often restricted by limited opportunities to become acquainted with them. We experience silverfish in unexpected encounters, when we remove a book from the shelf or pick up a box in the basement. The meeting is happenstance and only momentary. We generally consider them uninvited guests in our homes. We may be aware that they feed on starchy materials and may damage books or get into household food. But on the other side of the ledger, consider their complex mating rituals. Reflect on the minute yet intricate body plan incorporating digestive system, nervous system, musculature, and respiratory system. Through a microscope, one marvels at the elaborate silverfish body with its multiple segments, silvery scales, and long, filamentous antennae. Yes, they trend toward alien. Silverfish molt, they breathe through their sides, have multifaceted compound eyes, and sense much of their world through their antennae. They may be alien to us, but those differences are another remarkable way of living in the world. They are differences that are emblematic of the variety of ways that life expands and flourishes, exploiting myriad strategies for success on the earth. And silverfish adaptations are what make them good at living with us.

Silverfish are simply one example of the nature that is woven into our daily activities but that we generally don't see. These plants and animals are too small or too common to register or are just part of the daily landscape. Perhaps the organisms that we see regularly around home and in our towns and cities lack the exotic appeal of those we find in wilder places. Yet for a silverfish, it's pretty wild here at home. They're busy interacting with us and the other organisms that occur in their environment, including predators like spiders or centipedes, finding safe places to reside and discovering sources of food and moisture. Supposing for a moment that you were a silverfish, can you imagine the fright of someone lifting up your home, that box you've been living behind since you were an egg? Or imagine the adrenaline rush you would experience when a centipede comes looking to make a meal of you. We are not attuned to the daily drama involving silverfish and other plants and animals in our environment. Yet, despite their more mundane occurrence, these organisms that we find underfoot have a value derived from hundreds of millions of years of evolution and the clever adaptations that make them successful in our world. The dwellers of the world that humans have created comprise a biological richness that surrounds us each day no matter where we are. They are plants and animals that have succeeded, if not because of human beings, at least along with humans. These Anthropocene species merit a closer look—a look not just indoors but outside our homes as well: outside, where we find the spectacularly successful dandelion (*Taraxacum* species); indoors, where the invaluable little fruit fly (*Drosophila melanogaster*) thrives; and again outside, where we become reacquainted with another long-time partner, crabgrass (*Digitaria* species).

I step out of the house and survey the front lawn, a scene in which dandelions are a prominent, unintended feature. My adult encounters with dandelions have been all about lawns. The experience devolves to providing the mechanical force that causes plastic wheels to rotate gears that dispense small, colored pellets, a combination of fertilizer and herbicide, from a fertilizer spreader onto the lawn. There is a kind of pendulum effect as the spreader moves to one end of the lawn and back again in successive rows, wheels bumping along the uneven surface and gears grinding against the rock-hard pellets. The pellets, once released, nestle deep into the thatch, awaiting moisture to release the chemicals. The target of those chemicals, several dandelions, already in bright yellow flower, rise above the grass as a reminder that I should have fertilized sooner. My pristine, evenly green carpet is interrupted by undesired diversity. The color scheme is all wrong. The dandelions are interlopers elbowing their way through the grass, shoving the blades out of the way. They're not the only non-grassy plant in the lawn, but they certainly are the boldest, the most noticeable. How does this aggressor return year after year despite herculean herbicidal efforts?

I have trudged behind the spreader more often than I would care to admit, peering intently for the faint impressions left by the wheels. I have wielded a spade against dandelions and their seemingly infinite taproot. And I'm always ambivalent about my attempts to produce a dandelion-free lawn. Why not leave a couple of dandelions? There is a house down the street that seems to have more dandelions than lawn. That might be a reason. But maybe it's time to let the lawn grow a little wilder. Recent research says that more insect pollinators are found in a lawn that is cut less often. More pollinators sounds appealing. Maybe letting the lawn go and allowing a few dandelion colonists would be like a deep exhale, a sort of landscaped letting my hair down. Maybe I could say that dandelions add interest to a front yard, evoking a devil-may-care curb appeal. Besides, there is a lot to like about dandelions if we can learn to get along.

Dandelions might be one of the most successful plants in the Anthropocene age, thriving where humans have made changes to the earth's surface. Grow where you're planted is an aphorism we use for living well in any given situation. Dandelions are a model of that flexibility. They can flourish in an astonishing variety of situations, pushing through cracks in the sidewalk, in vacant lots, and on gravelly roadsides. Dandelions owe their success to multiple evolutionary adaptations that become apparent as we learn about their life cycle.

Picture the dandelions out in force. Some of them have developed full yellow flowers and others have already reached the seed stage. Late in the afternoon, big, curvy cumulus clouds drift across the deep, summer sky. A little girl on her way home darts across the yard, stops midway, and stoops to pluck a dandelion stem. She draws a deep breath and blows across the seed head, scattering the fluffy, tiny seed parachutes into the breeze.

Let's say you are the dandelion seed and are so light that a faint updraft carries you high up toward the clouds. Rising above the treetops, you catch a thermal wind gust and are drawn even higher. The seed flight is exhilarating. You can travel for miles up here. It's one day, maybe two, and you hit a downdraft that drives you earthward. The wind slackens as you near the ground and you float down, dropping between blades of grass in the center of a lawn. You settle on the soil surface to await the moisture that will start the growth of a new dandelion.

For a time, the parachute remains attached, and, when jostled by the breeze, rocking back and forth, it helps work the seed into the soil. The needed moisture comes in the form of a light rain. The seed rapidly germinates and spreads a shallow system of fine roots to obtain moisture and nutrients. The lawn is tidy, cut close to the ground, allowing the sun to warm the soil surface and hasten the germination and rooting process.

As you grow into a plant, you are sending a taproot deep into the soil. Your taproot grows ever deeper, maybe as deep as fifteen feet below the surface, to search for sources of moisture and nutrients. In the process, you're improving the soil for other plants as you aerate it and draw up minerals inaccessible to more shallowly rooted plants. Should someone happen along and attempt to uproot you, most likely a portion of your taproot will be left in the ground and you will quickly regenerate into two or three new plants. But let's say instead that you're left alone. Undisturbed, you send out a set of leaves just above the soil surface in rotating fashion. Your leaves elbow aside the nearby blades of grass. The flattened aspect of the leaves at the soil surface prevents other plants from approaching you, and as the leaves spread out you gain control over the sunshine and moisture that falls in the enlarging area. Leaves that nestle close to the ground are also less likely to be eaten by grazers and damaged by the elements than those of plants of greater stature. The edges of your leaves tilt downward toward the midrib at their center and drain fallen moisture toward the middle of the plant and your roots below. You have developed an amazingly efficient structure for growth!

All of that growth would be wasted, though, without an equally marvelous system of reproduction. You, the dandelion, extend a long hollow stem carrying the flower head above the leaves. The flower head is not one but many individual flowers. Each of what appear to be "petals" is a flower. Every tiny flower produces a seed, so that a single flower head may have as many as three hundred seeds. All of a sudden, you're a reproductive machine. Leafy green bracts surround the flower head and protect the flower's precious store of nectar from crawling insects that will not efficiently distribute the flower's pollen. Dandelion nectar is a great source of food for pollinators that fly, like bees. Despite all of the nectar you provide, like other dandelions in North America, you don't need pollinators. You can form seeds on your own without fertilization. Above the bracts, your green sepals envelop the flower in a budlike form at night or on rainy days. Finally, you close your sepals a last time to allow the seeds and parachutes to form inside. The flower stem sags and your head rests near the ground. When seeds and parachutes are ready, the flower stem stands tall again and the sepals open to reveal your white, spherical seed head. It's an early summer day and the little girl plucks the seed head, draws a deep breath, and blows across it. Dandelion and human descendants rise triumphant in an Anthropocene world, a world conformed both to human and dandelion needs.

Returning indoors to the kitchen, I wave my hand over a fruit bowl and a cloud of small flies rises into the air and disperses around the kitchen. These small insects are often called fruit flies, though other, quite different flies share that common name. Fruit flies appear in my kitchen through a process of what seems to be spontaneous generation. Their sudden appearance and rapidly increasing numbers can be disconcerting. We're competing for the same fruit and I'm outnumbered. Most likely, when I bought the fruit at the grocery store the fruit flies came along for the ride. In that sense, maybe they do have some kind of first finder's rights to the fruit.

Normally, I have a live-and-let-live arrangement with fruit flies. But occasionally they are numerous enough, rising up like a small squadron as I reach for a banana, that I must act. I may deploy a bowl of vinegar (moderately successful) or try to swat them in the air (rarely successful). Ultimately, though, I only prevail if I eat all the fruit and dispose of the pits, rinds, and peels before the fruit flies become thoroughly established and make themselves at home in my kitchen. Tolerance of fruit flies varies among people. Some of the less tolerant deploy sticky flypaper and fly swatters against fruit flies. Others, with a bent toward higher technology, may turn to vacuums and hair dryers to eradicate them.

Though they may occasionally be troublesome, fruit flies have their attractions. In a roundabout way, fruit flies are why I entered the field of entomology in the first place. I first encountered fruit flies while working as a laboratory instructor, using them for experiments to demonstrate genetics to nursing students. Fruit flies are ideal laboratory subjects. They are easy to grow and have impressive reproduction capabilities (each female lays 400–500 eggs). Examining flies under the microscope, I was fascinated by the minute detail of their honey-colored bodies, their brilliant red eyes, and their small antennae ending in a feathery tip. In their experiments, students recorded the number of mutations in eye color, body color, and wing type. Will the body be ebony or tan? Will the eyes be brown or red? Will the wings be crinkled or normal? Calculating the ratio of offspring that exhibit each of these characteristics—say three tan flies to one ebony fly—allowed students to determine that the tan fly body is genetically dominant whereas the black body gene is recessive. Beyond the exercise itself, however, students experienced the thrill of witnessing first hand how living creatures are organized. But fruit fly contributions to biology are not limited to genetics. The ubiquitous little fruit fly has been the subject of studies that have significantly advanced the fields of ecology, animal behavior, and evolutionary biology. Fruit fly studies have opened a door to the inner workings of nature for me and many other students of biology. As I learned more, I realized that a lot of fascinating biological work was being done not just with fruit flies, but also with many insects in the laboratory and outdoors. And, that led to a lifetime interest and a new career.

A lot more is going on in a fruit fly's short life than a casual observer might suspect. Take sex. Back in the fruit bowl a male fruit fly has just discovered a lonely female. It's a pleasant evening, and they luxuriate in the agreeable aroma of ripening bananas. He walks up and taps her tentatively with his foreleg. His foreleg senses chemicals on the female's exterior that help him determine whether she is the right species and a mature female. Next, he unfolds a wing and vibrates it in front of her head. The vibrating wing is producing a love song composed of pulses and humming. He walks around in circles in front of or behind her but always facing her. From time to time he may switch wings to ensure that he's vibrating the wing closest to her head. At this point, I need to warn you that fruit flies are not nearly as demure as silverfish. He proceeds to her tail, and attempts to mount her from behind. But she has the last word in the process, for she must accept his advances for him to be successful. Thus, if she is not satisfied with his dancing and fanning, he will leave disappointed, in search of another. The female is in the evolutionary driver's seat, and the selectivity of countless generations of female fruit flies has led to the peculiar courtship display of the male. Some biologists speculate that over evolutionary time human females have selected for human male characteristics too.

Fruit flies appear to be rather peaceable creatures, but both sex and food may result in battles on the banana peel. Males tend to be more aggressive than females, fighting over both mates and food. Females only fight about food, not mates. I guess fruit fly males aren't worth the trouble. Males have quite a range of aggressive behaviors, including threatening by raising their wings toward their foe or, if that's not sufficient, physically attacking each other. Male fruit flies have been observed standing upright and boxing with their forelegs and may also engage in behaviors that scientists refer to as tussling, kicking, holding, or chasing. Coming between a fruit fly and his intended can be a hazardous affair. Females may engage in many of the same aggressive behaviors but are somewhat more restrained. Fruit fly females don't box and certainly don't tussle.

Fruit flies have been used to answer all sorts of biological questions, from exploring how to extend life spans to understanding how neurodegenerative diseases affect the brain. Fruit flies are most well known for their use as models for basic studies of genetics and evolution. But I was surprised to learn that they also have been subjects for experiments on the effects of alcohol. Fruit flies naturally feed in fermenting fruit and are often exposed to high levels of alcohol. Fruit fly young have even been found living in fermenting liquids such as cider, wine, and beer. Thus, we have learned that fruit flies can efficiently process alcohol and are resistant to alcohol poisoning. Despite fruit fly alcohol resistance, when scientists cause experimental intoxication of fruit flies they observe the same kind of stimulant and depressant effects found in humans. Intoxication changes fruit fly social behavior just as intoxication may affect human social behavior. Intoxication reduces male fruit fly inhibitions to the point that males are inclined to court the next partner they see. But do fruit flies suffer remorse the next morning?

Scientists have also been studying how fruit flies age, since age-related decline in fruit flies bears striking similarity to aging in humans. The fruit fly life span is only about fifty to eighty days long, and deterioration in capabilities begins to take place fairly quickly—in as little as ten days. Fruit flies have to cram a lot of living into those first ten days. As they age, flies don't get around as well and don't explore as much; their sense of smell declines; they don't get as much rest at night; they don't remember as well; and, yes, the percentage of males that copulate decreases with age and females lay fewer eggs. This all sounds hauntingly familiar. Happily, scientists have been able to both increase life span and delay functional decline in fruit flies. Could humans be next?

Although fruit flies have been eating, drinking, and making merry far longer than humans, human beings now apparently provide the main means for global fruit fly revelry. Fruit fly ancestors developed about 100 million years ago. At that time, flowering plants were emerging and dinosaurs were dominant. Fruit fly ancestral young originally fed on rotting vegetation but later evolved to feed on yeast and bacteria associated with fruit, tree sap, or fungi. Recently, in evolutionary terms, perhaps in the past ten thousand years, fruit flies have evolved to associate with humans. An intriguing paper published in 2018 speculates that this association arose in sub-Saharan Africa, when humans brought the fly's preferred fruit from the morula tree (*Sclerocarya birrea*) into their caves.[1] Today, we find fruit flies living primarily in association with human activity or habitation. Scientists suggest that fruit flies now require the continuously available food sources that humans provide rather than the less reliable resources available in non-human nature. The connection with humans has led to a cosmopolitan distribution of fruit flies, hovering over fruit bowls from New York to New Delhi. Our fates are inextricably entwined. We depend on fruit flies to develop increasingly deep insights into humanity, and fruit flies depend on us to provide food and habitat. Fruit flies have become part of our Anthropocene family, thriving on our fruitful excess.

I leave the kitchen for my backyard, where the uniformity of the lawn is disturbed by clumps of lighter green plants that seem all stems and broad grass blades. These homely clumps are sources of food in many parts of the world, although they are now considered a weed in North American yards. It's crabgrass, which includes multiple members of the plant genus *Digitaria*. Several species of *Digitaria* have been sources of human food, as millet, from early agriculture until today in China, India, and Africa. Central European immigrants in North America also valued crabgrass seed as food, and it is sometimes called Polish millet. Crabgrass originated in Eurasia and was imported into Europe and North America primarily as animal forage. However, its qualities as food for grazing animals and millet for humans have subsequently been overshadowed by its invasion of yards and gardens. What was once prized is now vilified and looms large in our North American suburban consciousness. Its iconic status in American yards is memorialized in book titles that refer to suburbia as a "Crabgrass Frontier" or a "Crabgrass Crucible."

A search of newspaper articles on crabgrass is revealing. The headlines exhaust warlike vocabulary in describing crabgrass-human interactions, speaking of killing, attacking, routing, exterminating, battling, fighting, and just plain getting rid of crabgrass. Crabgrass is a "nemesis" that produces "tension" and "neurosis," according to the *New York Times*. So it's not surprising that crabgrass is a significant target for herbicide use. But the scrappy crabgrass plant fights back. Crabgrass has developed resistance to multiple herbicides. Furthermore, the plant produces a few chemicals of its own. These chemicals, generated by the crabgrass roots, alter the microbial community in the soil and inhibit the growth of other plants. Although I've tried my hand at uprooting crabgrass, not a happy task, I find myself wishing these plants success. There is something about their robust, I'll-take-what-you-give-me nature that stands in stark contrast to the effete bluegrass lawns that require constant attention.

Crabgrass employs a more efficient system of photosynthesis than most other plants, and it grows faster than most turf grasses. The metabolic efficiency permits crabgrass to survive well in conditions of drought and poor nutrition. In fact, crabgrass prefers hot, sunny locations. Crabgrass is an annual—the life cycle of an individual plant lasts a year. A local earthworm would observe seeds starting to germinate in the early spring when soil temperatures reach about fifty-five degrees Fahrenheit. She would see a single shoot emerge from the seed. As the summer season progresses, the crabgrass plant produces more leaves and also stems growing laterally from the main stem that are called tillers. A single crabgrass plant can produce up to seven hundred tillers. The earthworm would detect roots forming where nodes of the tillers touch the ground. The plant itself can grow as tall as two feet. Growth slows later in the summer, when the stems produce delicate fingerlike projections on which tiny flowers called florets are arrayed. The flowers are wind pollinated, and a single crabgrass plant can produce as many as 150,000 seeds. The seeds may be distributed by wind, attachment to the coat of passing animals, or in scat from grazing animals.

It is paradoxical that a plant that sustains both human beings and their livestock has become a suburban pariah. Crabgrass has attributes that make it a resounding success in landscapes altered by humans. In particular, the ability to grow in disturbed sites under sunny conditions with poor soil quality allows crabgrass to succeed where many other plants fail. It is another Anthropocene success story. Though we generally do not wish crabgrass well, we create conditions that allow it to flourish. An ambivalent relationship seems to be characteristic of human interactions with the organisms that thrive in our environment. Are they too common? Do we prefer to choose which plants and animals reside with us?

I return indoors to my bookshelf, which I seem to be sharing with the silverfish. Here I find that Aristotle, the father of biology, contends, "we should venture on the study of every kind of animal without distaste; for each and all will reveal to us something natural and something beautiful."[2] Nature, whether it is to our liking or not, is where we find it. If we look closely, nature will reveal something beautiful in the most unexpected places. It may be at home, in a yard, or in a crack in the sidewalk. The roadside and a vacant lot are paradigmatic environments in the Anthropocene. It is here that we find plants and animals that are well known to us but are unexpectedly beautiful—beauty not simply in appearance but also found in the appreciation of multiple adaptations that have led to evolutionary success.

It's not readily apparent when you see them that silverfish

have an elaborate mating ritual or that they could learn to navigate a complicated maze. Dandelions can definitely be an irritation in lawns and flower beds, but you have to admire how their beautiful, floating seeds, long taproot, and rosette leaf arrangement equal success in human habitats. Who would think that the tiny fruit fly might be so similar to humans and also might provide critical scientific information that enables us to practice better medicine? And what of crabgrass, an ancient human food source that has followed us around the planet and lives unbidden in continued close connection to us?

Each of these amazing organisms belongs to the mundane part of our lives. They are what we see every day at home or on the way to school or work. They are seen but unseen. They are commonplace because they are the winners in our epoch. They are nature underfoot—found on and around the paths we travel every day. These plants, insects, and other organisms are common and often considered undesirable. Yet they present an ideal opportunity to find wonder in nature. The prevalence of these species provides a chance to get to know them well and to appreciate the evolutionary history inscribed in their genetic material. That history creates unique capabilities and behaviors. These are lives that can provide enchantment just like the more charismatic organisms found in national parks, zoos, or botanical gardens. Silverfish, dandelions, fruit flies, and crabgrass are but a microcosm of the communities of plants, animals, and microorganisms that we create around us in the Anthropocene.

We're up against them, and they're living right up against us. We are cheek by jowl, or perhaps cheek by antenna, with a nature of our own making. It is wild nature, pursuing its own ends in our homes and yards. It is wild nature in our basement, where two silverfish meet, at the intersection of damp cement floor and concrete block wall, shyly touching antennae.

TWO

Nature at Work

THE JANUARY SUNSET is disappearing in golden hues on the South Carolina horizon as she flies back to the hive carrying her last load of pollen and nectar. It's a placid scene, with no indication of the dislocations yet to come. The field honey bee (*Apis mellifera*) is one of the last to return at the end of a long day of foraging. On entering the hive, house bees approach her and collect the pollen and nectar she gathered. Pollen and honey supplies were ample in the fall, and the colony has remained strong through the winter months. But colony activities are suddenly disrupted as smoke wafts into the hive. The smoke makes it difficult to sense whether the guard bees have deposited alarm pheromone, which would signal the bees to defend the hive. Is the colony in danger? The field bee and those around her quickly ingest a stomach full of honey in case a fire threatens the hive. As a result, they become more docile. The source of the smoke, rather than a fire, is a beekeeper. Beekeepers use smoke to quiet the bees before disturbing the hive. In this case, they are preparing the hive for a move.

Moments later, the hive rocks gently as it is lifted by forklift onto a flatbed truck. It will be transported to a holding yard in preparation for a much longer journey. The motion of the truck is disorienting, and perhaps stressful to the bees. Fortunately, the trip to the holding yard is short, although the holding yard is only the first step in a monumental migration for the honey bees. The bees adapt to their new location, making short orienting flights, and then longer foraging flights. Several days later, shortly after the last bees are returning in the evening, beekeeper-generated smoke again wafts into the hive. The reaction is the same. The bees quickly ingest honey and are docile yet alert to sense alarm pheromone dispersed from the hive's guard bees. The alarm never comes—the bees can't distinguish the pheromone from the smoke.

Again, the hive rocks as it is lifted onto another flatbed truck trailer. A net that prevents the bees from escaping is stretched over the hives and closed with staples that reverberate in the hive boxes as they are driven into the wood. From inside the hive, the bees can feel the ratcheting vibrations of the buckles as heavy straps are pulled tight across the hives to secure them to the truck bed. The loading goes on well into the night as other trucks are stacked with honey bee hives. Finally, the trucks are full, and more than a thousand hives are ready for a trip of several thousand miles to almond orchards in California.

The trucks start up and diesel exhaust envelops the holding yard and the hives. Slowly, the trucks pull out and begin their journey. The trucks lumber across the southern tier of states on their way to California. The bees inside the hive feel a steady vibration and jostling as the trucks roll down the highway. None of them have experienced anything like this except perhaps the queen, who generally outlives the workers. Along the way, a late January cold front reaches down into Texas, causing temperatures to drop into the thirties as the trucks pass through. The cold causes the bees to cluster together to maintain temperatures in the hive. As the trucks cross the desert in Arizona, temperatures rise into the upper eighties. Stopping here could be fatal for the bees, since they can't escape the overheated hives. The truckers douse the hives with water at a truck wash to protect the bees from the heat. The bees fan their wings to evaporate the water and cool the hive. Having survived weather extremes and long days on the road, the trucks come to a stop at the California border, where inspectors search them for unwanted hitchhikers, such as ants or weeds. No pests are found on the trucks, and ultimately, the hives arrive in the almond orchards of central California.

After the bee net is removed, countless bees make orienting flights and cleanse themselves after the long journey. The blossoms have just appeared on the almond trees and a sea of blooming white stretches as far as the eye can see. There are 750,000 acres of almonds in California, and 1.5 million honey bee hives are required to pollinate them. That represents more than half of the hives in the United States—used solely to pollinate almond trees in the spring. The bees are not there to produce honey, since almond honey tends to be rather bitter. The bees are employed only for pollination. When almond pollination is completed, the human-assisted migration of honey bees will continue for many more thousands of miles through the summer— perhaps to cherry trees and apple trees in the Pacific Northwest, to blueberries in the east, and maybe pumpkins in Pennsylvania.

The bees may finally end up in New York or Michigan or Montana in the late summer to finally focus on honey production.

Humans have transported honey bees for millennia. What has changed in the Anthropocene is the distances bees travel, crossing multiple time zones on trips that last for days. Transport of honey bees has become critical to the pollination of crops, because many native, naturally occurring pollinators have declined or disappeared. The numbers of native pollinators, made up of many species of bees, moths, and flies, have declined. This is primarily a result of habitat loss and pesticide use. A fatal syndrome, called colony collapse disorder, which is likely a result of multiple factors, has also recently affected honey bees themselves. Pesticides, disease, parasitic mites, and perhaps the stress of long-distance travel may all be contributing to honey bee colony collapse. The loss of local pollinators combined with threats to honey bees has become a global risk for our food supply.

I raised bees for honey on a friend's farm for a few years. There were just a few hives kept on a small, hobbyist scale—nothing like the big honey bee operations that ship bees all over the country. Raising honey bees is a rewarding activity—partly for the bounty of excess honey, but also for a connection to the bees themselves. Maintaining successful bee colonies requires learning about what threatens bee survival, and what allows them to flourish. For example, a beekeeper should consider an optimal location for the hive that is warmed by the sun early in the morning and shaded during the hottest part of the day. Beekeepers in cool climates must be sure that bees are well provisioned with honey for the winter months. Honey bees must also be watched for diseases and parasitic mites. There is much to learn for the amateur beekeeper, making the production of honey all the sweeter.

Pollinators are declining partly because of the changes that humans make to the landscape. In modern agriculture, large swaths of land are often devoted to a single crop plant, reducing or eliminating the pollinators that specialize in feeding on specific wild flowers. Generalist pollinators like many bumble bees (genus *Bombus*) and honey bees, which visit a wide variety of flowers, adapt best to these circumstances. Further, when a single crop covers a large area, like almonds in California, local pollinators find it difficult to collect sufficient nectar during much of the summer. This is because a single flush of flowers occurs at one time across the almond orchard, and after that few nectar resources are available to native bees over the many acres covered by the orchard. As a result, pollinators must find unconnected patches of suitable flowers, sometimes separated by long distances. Long-distance travel between flower clumps favors pollinators that are stronger fliers, like bumble bees. On the whole, our industrial-scale agricultural practices, while efficiently using land, have resulted in the absence of local pollinators—and that drives the honey bee migration business. Although it is effective for pollination, transporting honey bee hives over long distances stresses bee colonies, and exposes them to a variety of pesticides, diseases, and parasites.

The complexities of pollination for industrial agriculture are just a part of the story of human dependence on small beings, whose importance may be overlooked. We are well aware of the range of plants and animals that we depend on for food. But we are also dependent on broad assemblages of tiny animals that for their size and ubiquity often escape our notice. These groups of animals are actively involved in a variety of tasks that assist us— from the beginning of life, as pollinators like the honey bee and bumble bee, to the end of life, as decomposers and recyclers such as millipedes (order Diplopoda) and hover flies (family Syrphidae). Hover flies, which often have yellow and black coloring and look a lot like bees or wasps, are important pollinators but can be decomposers too. These stinging-insect mimics operate at either end of the spectrum, both pollinating and decomposing. Anthropocene habitats alter the composition of this assemblage of organisms so necessary to humans, favoring some and eliminating others. Bumble bees, although in decline in many places, are pollinators well adapted to our changing landscape. Fuzzy and lovable (to some), bumble bees create amazing societies. Millipedes are decomposers that we can find occasionally indoors and in our gardens. This many-legged animal is in one way or another processing our wastes. Some of these animals may succeed in the new Anthropocene landscape, benefiting from the changes we make, but others are less successful, which creates a concern because we require their services.

When I was young, my father claimed that he used to pet bumble bees when he was growing up. I never summoned the courage to try, but bumble bees became an object of fascination for me. Bumble bees have robust, hairy, black and yellow bodies. They do look soft and pettable. Their sting, though, explains my reluctance to pet them. Unlike honey bees, whose stinger is barbed and remains in the skin, the bumble bee stinger is smooth, allowing them to sting multiple times. The honey bee sting is a suicide mission. The bees die as they pull away from the stinger, leaving entrails, and a pulsing poison sac attached.

I don't know if my father ever really petted bumble bees, but he did have something to do with a later turn toward studying insects. He gave me a butterfly net when I was four or five years old, which I used haphazardly, chasing swallowtail (genus *Papilio*) and monarch (*Danaus plexippus*) butterflies in our neighborhood. I don't remember catching many, but simply having the net and searching for butterflies caused me to see my surroundings in a different way—populated by desirable fluttering beings. I wouldn't recommend a butterfly net for a five-year-old today, given the potential to damage or kill these beautiful and sometimes endangered insects, but searching for and naming butterflies at any age may open a new way of seeing nature. My father also purchased an ant farm for me at around the same

time. In the farm, a small colony of ants toiled in a sandy medium, narrowly compressed between clear plastic panes. We
could see the ants tunneling through the sand and transporting
bits of food we dropped on the surface. My father's efforts at
developing an appreciation for other organisms weren't limited
to insects. Goldfish, tropical fish, and small turtles were housed
in various aquaria in my room extending into my middle school
years. His last contribution to insect awareness was a field guide
to the insects of North America that he gave me in high school,
during a long summer spent working at my grandparents' ranch
in southeastern Oklahoma. The glossy pages with color illustrations of many exotic-looking creatures were fascinating, and I
made some limited attempts to identify insects I found around the
ranch, including a beetle that may have been an American burying beetle (*Nicrophorus americanus*)—discussed later in this book.
My father's efforts didn't immediately take hold, as I retained
only a mild and sporadic interest in insects. He did, however,
cause me to notice insects as legitimate objects of curiosity—
worthy of deeper consideration.

Like honey bees, bumble bees are very important pollinators,
in part because most bumble bee species visit a variety of different flowers and can travel long distances when those flowers are
widely dispersed. In many respects, bumble bees are better pollinators than honey bees. The bumble bee's large size and hairy
body results in better contact with pollen-bearing parts of flowers and better transport of that pollen. The longer tongue of most

bumble bees also contributes to better pollination, particularly in certain plants like beans and peas. Bumble bees also engage in buzz pollination, in which they hold tightly to flower parts and vibrate their flight muscles. This causes tightly held pollen to be released in the flower. Buzz pollination is important for plants like tomatoes, kiwi fruit, and blueberries. Bumble bees are faster pollinators than honey bees. They can pollinate more flowers per individual bee in a given period of time. However, honey bee colonies, averaging 20,000 to 50,000 workers, are much larger than those of bumble bees, which average about 40 to 120 workers. So honey bees' colony size makes up for their lower efficiency in pollinating many crops. Bumble bees are also more tolerant than honey bees of less than optimal environmental conditions. They can fly and pollinate flowers at forty-one degrees Fahrenheit, whereas honey bees will not leave their hive until temperatures rise above fifty degrees. Bumble bees can fly at lower temperatures because they use their flight muscles to shiver, thus warming their bodies enough to begin to fly. The ability to fly at lower temperatures makes bumble bees important pollinators of plants that flower at cooler times of the year, such as in the early spring, and enables them to start their day a little earlier than honey bees.

Bumble bees are preferred pollinators for a number of agricultural crops. They are very effective pollinators of greenhouse-grown crops, such as tomatoes. Honey bees tend to become disoriented in greenhouse situations, but the enclosed environment doesn't seem to trouble bumble bees. In fact, bumble bees are now commercially produced and sold in container hives to farmers for greenhouse and outdoor pollination around the world. Unfortunately, the commercial use of bumble bees appears to have had some negative consequences. Commercially reared bumble bees are reported to have spread parasites and diseases to naturally occurring pollinator bees. The occurrence of these parasites and diseases has also been implicated in the decline of the other pollinator bees.

Some bumble bee species remain abundant in the Anthropocene, but many bumble bee species are present only in reduced ranges and smaller numbers. A variety of factors, including the aforementioned parasites and diseases, have led to the decline in bumble bees, and even the extinction of some species. Intensified agriculture, and development of land for other human uses, decreases the number and diversity of flowering plants and has contributed to the decrease in bumble bee populations. The widespread use of pesticides has also likely impacted bumble bees. Concern about the decline of bumble bees and other pollinators has led to several movements in North America and Europe that are dedicated to providing floral resources for bumble bees and other pollinators. Growers are also recognizing the need to set aside areas at the edges of farm fields where local wildflowers can grow and support native pollinators.

Like honey bees, bumble bees are social insects, with multiple generations of workers living together and performing specific roles. Bumble bee societies differ from honey bees in part because the colony lasts only a year. Honey bee colonies are perennial—so in the absence of external disturbance, and with the ongoing replacement of queens, honey bee colonies can continue for many years. Honey bee queens themselves live for three to four years. Most bumble bee colonies, however, die at the end of the season. So in one summer, the bumble bee colony develops from a single insect (the queen) to a society functioning with multiple coordinated roles. In the process, the bumble bee colony goes through several distinct phases during the year, focusing first on colony growth and then later on reproduction. It all starts with the queen, who, after mating in the fall, spends the winter in an underground cavity. She emerges in the spring to feed and begins to lay eggs in wax cells, often starting the nest in an abandoned rodent burrow. This is the beginning of the solitary phase of the colony, in which the busy queen performs all of the tasks in the nest. The queen alone literally does it all: foraging for pollen and nectar, caring for the brood, and continuing to lay eggs in the nest. The emergence of the first generation of workers begins a cooperative phase, in which the workers assume responsibility for foraging and brood care, and the queen for laying eggs. This must lift quite a large burden from the queen, and one can imagine her unwinding in the nest after a busy spring, cared for by an attentive brood of first-generation workers.

Later in the season, as the numbers in the colony grow, a chaotic competitive phase ensues. Some workers become aggressive, and a dominance hierarchy forms within the colony. This hierarchy is formed through antagonistic interactions between bees. Common aggressive actions include humming, or vibrating wings toward another bee, or darting, which involves a movement in the direction of a nearby bee. Or, the workers may physically engage by dragging a bee, biting it, or attempting to sting. In order to avoid these aggressive interactions, mild-mannered workers may actually produce a scent that informs aggressive bees that they are sterile, and thus noncompetitive. This allows the sterile workers to go about their business without interference from the troublesome aggressive bees. The successful dominant, aggressive workers, which are not sterile, may then begin to lay eggs themselves. Because the workers are unfertilized, the eggs only develop into male bees—female workers or new queens develop only from the fertilized eggs laid by the queen. Often, the queen, or other workers, will eat the eggs laid by the dominant workers. During this period of chaos, which occurs toward the end of the life of the colony, males and then new queens begin to emerge. Males contribute little to the colony and leave the nest rather quickly. The new queens must stay in the colony longer, provisioning themselves with food stores so that they can survive through the winter after mating. These queens are the bridge from the old colony to next summer's colony, carrying the eggs that create a thriving society in just a few months' time.

There are about 250 species of bumble bees worldwide, and

their life histories all vary to some degree from the colony described above. A very different, non-pollinating example is the cuckoo bumble bee (subgenus *Psithyrus*). The cuckoo bumble bees do not form their own colonies but use those formed by other bumble bees. They are parasites in bumble bee society. Cuckoo bumble bees do not produce worker bees and fully depend on their host colony to rear their young. Species of cuckoo bumble bees use several strategies to invade a nest without being killed by the colony workers. They can use physical dominance, in a sort of home invasion, which may include killing the host queen and some of the workers. Other, slyer cuckoo bumble bees can develop or acquire the scent of host bees in the colony and so seem to be native to the nest. Cuckoo bumble bees may also exude a repellent chemical that keeps the colony workers away. Once grown, the young, emerging cuckoo bumble bees are also able to mimic the chemicals of colony worker bees, which would otherwise kill them. Cuckoo bumble bees are masters of deceptive chemical communication, and that enables their success as social parasites, stealthily taking advantage of their bumble bee hosts.

Bumble bees in general communicate a lot of information chemically. Not only do they confirm identity and sterility or fertility by chemicals, but they also inform the colony of food resources, determine that a flower has been visited, and males advertise their presence to females via chemical pheromones. The evolutionary success of bumble bees is tied, in part, to the development of effective social networks, and the chemical communication that supports them. The instances of bumble bee success in the Anthropocene are related to their flexibility in accepting a variety of flowers, and ability to utilize patchy floral resources. For many bumble bees that's not enough, but I am hopeful that efforts to provide more floral resources for them in backyards and on the edges of agricultural fields will support their buzz pollination well into the future.

Bees are not the only pollinators. Other animals such as birds, bats, butterflies, moths, and flies also play important pollination roles for a variety of plants. The hover fly is probably the most important fly pollinator. There are about six thousand species of hover flies around the world, and they vary considerably in appearance and life history. If you are uncertain about what a hover fly might look like, think of small flies hanging in the air in an almost stationary position above a flower garden, and looking a lot like a bee or a wasp. Hover flies often have yellow and black coloration and body shapes that immediately suggest that they might sting you. And, of course, that's their strategy. Over evolutionary time, they have adopted the body plan and coloration of stinging bees and wasps in order to deter predators. When held, some hover flies will even press the tip of their abdomen against your finger as though really stinging, presumably in an effort to startle so that you'll release them. Hover flies mimicking wasps will even wave their legs in front of their heads to simulate a wasp's long antennae. Other hover flies imitate bumble bees with large hairy bodies and matching coloration. Despite the similar appearance, hover flies are quite different from the insects that they mimic. They are not social insects. They do not sting. They have two wings rather than four. And their youthful larval or maggot stage occurs in circumstances very different from the nest of a bee or a wasp.

I have always been enchanted by both the mimicry of hover flies and their flight capabilities. Like all true flies, hover flies have a single pair of wings and a pair of balancing organs, called halteres, which act as a sort of gyroscope, balancing them in the air. Hover flies display incredible flying ability, performing feats that no human-built aircraft could ever achieve. They are also one of the few insects that can fly in reverse. Hover flies can switch from hovering to darting off in pursuit of a potential mate in a millisecond. But much of their time is spent hovering above or alighting on flowers in our gardens. Just how hover flies achieve their aerobatics has been thoroughly studied by scientists and engineers wishing to apply lessons from hover flies to mechanical vehicles.

Some hover flies have dual value for agricultural production. Adult flies visit flowers and pollinate them, whereas young flies, in the larval or maggot stage, are efficient predators on the aphids that harm agricultural crops. The adult female hover flies are adept at locating aphid colonies on plants. It seems that they can also distinguish groups of aphids that are parasitized and avoid laying eggs near them. Scientists have experimented with planting flowers like alyssum, which hover flies prefer, at the edge of agricultural fields. The presence of these flowers both increased the numbers of hover flies and, because of the predaceous hover fly larvae, reduced the numbers of aphids on adjacent crops. In this case, adding flowers that attract hover flies could reduce or eliminate the need for the use of insecticides.

The young of other hover fly species play a role in the decomposition of organic material. The unappealingly named rat-tailed maggot, the larval stage of a hover fly called the drone fly (*Eristalis tenax*), is found in stagnant water, carcasses, or animal waste where it feeds on decaying organic matter. The rat-tailed maggot gets its name from a tube that is between one and two inches long, extending from the tail end of the larva. This tube is a breathing mechanism, something like a snorkel. It allows the larva to remain submerged headfirst in the stagnant water or waste in which it feeds. The drone fly was introduced to North America in the late nineteenth century. The adult fly is orange to brown in color with black markings and resembles a honey bee. This may explain surprising references from the Roman poet Virgil and in the Bible in which bees are obtained from an animal carcass. Of course, these "bees" were most likely drone flies whose immature stages fed in the carcass. The adult drone fly, like other hover flies, is also an important pollinator. Insects that play a decomposing role, like the rat-tailed maggot, act as an intermediate stage in the decomposition of organic matter. These insects break down materials into a form more readily utilized by bacteria, the final decomposers. Without these decomposers, we would be drowning in plant and animal waste materials.

Other hover fly larvae may be found burrowing in decaying wood or in rotting tree holes. They are also breaking down dead or dying materials as they participate in returning the tree and its nutrients to the soil. The presence of these specific kinds of hover flies is an indicator of a healthy old-growth forest, since this is where downed dead trees and snags (standing dead trees) may be found in abundance. Alteration of the forest that eliminates dead trees has caused the loss of these hover fly species in local areas.

Hover flies present a fascinating study in evolution both in larval food preferences and in adult mimicry of stinging insects. The adults of the various species of hover flies generally have similar lives that are associated with flowers. But, as we have seen, the immature or larval stages of hover flies vary widely in their food and environmental preferences, ranging from plant eaters to dead-wood feeders to filter feeders in stagnant water to predators of other insects. It is thought that the earliest ancestral hover fly larvae fed on fungi or bacteria. Plant feeding seems to have followed, with predation on insects and feeding on decomposing materials coming after that. Though possessing somewhat similar environmental preferences, the adults vary considerably in the degree to which they mimic wasp or bee models.

When scientists have reviewed the range of hover fly mimics in relation to their stinging models, they find two surprising results. First, there are many hover flies that are poor mimics, meaning they don't resemble their bee or wasp model very well. Shouldn't birds and other predators be able to discern the poor mimicry and eat them with abandon? Second, there are many more mimics than models. Does this lessen the value of mimicry in preventing predation, since the risk of a sting is lower? There is a great deal of scientific speculation on these questions. An interesting line of thought relates to the noxiousness of the model insect to predators. It is thought that bumble bees are the least noxious of the stinging insects to bird predators, though their big hairy bodies require a great deal of time and effort to eat. Wasps are the most noxious, with honey bees falling in between. Scientists suggest that the least noxious models require better mimicry, and the more noxious models (the wasps), representing greater risk to the predator, require less. We do seem to find that hover flies mimicking bumble bees tend to be more accurate mimics than the hover flies mimicking wasps.[1] Perhaps being close enough in appearance to something really distasteful is sufficient to deter bird predators. In any case, I'm glad of their successful mimicry since these little flies and their immature stages provide multiple benefits to humans and the ecosystems in which they occur—and they also provide a great deal of pleasure when I see them hovering around plants in the garden.

The hover fly larvae that feed in dead trees are but one part of a larger community of decomposer organisms. Some ancestral hover fly larvae can be found feeding on bacteria and fungi in leaf litter on the forest floor, where another important member of the decomposer community, the millipedes, may also be found. Millipedes, like insects, are joint-legged creatures. Whereas adult insects have six legs, though, millipedes have many more. Adult millipedes may have from thirty legs to over seven hundred, depending on the species. No one has found a millipede that is true to the name of this class of organisms, however—millipede is literally translated as a thousand feet. There are many species of millipedes, and their body plans vary, but they consistently have body segments with two pairs of legs. Millipedes aren't born with so many segments and legs, though. When they emerge from their egg, young millipedes have only a few segments and legs. They add additional legs and segments with each molt when they shed their skin. This is a process that may take from one to five years, depending on the species, and generally consists of seven molts. Millipedes continue to elongate and sprout legs until they reach adulthood, living as long as eight to ten years.

Millipedes are an ancient group. The first air-breathing land animal was a millipede, living in what is now Scotland over 400 million years ago. This earliest millipede was only about a half-inch long. Later, though, other ancient millipede species grew to over eight feet in length. Today, the largest millipedes grow to about fifteen inches, but most are much smaller. Millipedes are found around the world in a variety of habitats, from the tip of Argentina to the Arctic Circle. So millipedes have been a successful group of animals for hundreds of millions of years. But, similar to bumble bees and hover flies, certain members of the group have done well in the Anthropocene, and others have become endangered. The endangered species seem mainly to be those that are specialized for living in particular habitats, like old-growth forests, that are more fragmented and limited in area now.

Millipedes are generally rather slow-moving animals that may often be found under logs and leaf litter. Their favorite hiding spots reflect their need for microenvironments of high humidity. Insects have an outer skeleton that contains a waxy layer to prevent moisture loss, but this waxy layer is not found in millipedes. Millipedes therefore seek out protected, moist areas. As a result, millipedes are often uncovered around human homes when raking up grass clippings or leaves, or in picking up a board that has been laying in the yard for some time. I often find them when I pick up a potted plant. Most millipedes feed on decaying plant material. It has been estimated that millipedes consume from 5 to 10 percent of the leaves on the forest floor in temperate environments. In tropical environments, millipedes may consume nearly half of the leaf litter.[2] Millipedes are very important decomposers, then. Occasionally, millipedes may invade homes, but it's not a good strategy for them. Indoor environments are generally too dry for millipedes to survive, and there is little for them to eat.

You might wonder how the slow-moving millipede can be so successful. How does it avoid predators? There are a few exceptions to the slow millipedes, including millipede species in Africa that can jump. But mostly millipedes rely on defenses other than speedy avoidance. Most millipedes are covered with armor-like plates on each segment that afford some protection. One millipede has even evolved spikes that emanate from the plates. Other millipedes take advantage of their armored protection and coil up, or roll into a ball, so that their head, legs, and unprotected undersides are not exposed. Certain very small millipedes have tiny hairs that they use to entangle predator ants. Millipedes are probably most well known for their range of chemical defenses, though. A few of the chemicals are repellents, while others burn and sting. Some millipedes produce hydrogen cyanide, which can be deadly in an enclosed area. These chemicals ooze out of most millipedes, but there are several species that have developed the ability to spray the chemicals when threatened. There is a record of a scientist who suffered blisters and temporarily lost vision in one eye when sprayed in the face by a millipede. More commonly, human encounters with millipede-generated chemicals result in what is termed mild pain and skin discoloration—the skin turns brown. There was a case in Oregon in which semicircular brown marks on a one-year-old child's stomach appeared to be burn marks to emergency room workers. The child was actually removed from its mother's care, despite her insistence that she had not harmed the child. It was only later that a millipede was found in the child's pajamas at her day care facility, and identified as the source of the brown marks.[3]

Since millipedes produce a lot of noxious chemicals, you would expect that they would advertise it with brightly colored bodies. Bright coloration is more the exception for millipedes though, which tend to be more brown or gray in color. Of course, most millipedes spend their time beneath leaves and logs, so perhaps there is little benefit in developing bright colors and more benefit in blending in to their surroundings. Some millipedes do sport bright reds, oranges, yellows, and even blue coloration. There are also several millipedes that produce bioluminescence—they glow in the dark. These millipedes are active at night, and presumably the bioluminescence warns predators to stay away. Field experiments conducted by scientists with luminescing and non-luminescing millipedes suggest this is true.[4]

Millipedes are quite an astounding evolutionary story. From their status as first animals to colonize land to the development of sophisticated chemical warfare against predators, millipedes have a 400 million year history of evolutionary success. From their beginnings on land they have consistently played a critical role in recycling energy and nutrients from plants. They are invaluable to the ecosystem, and therefore to us. The same is true of honey bees, bumble bees, millipedes, and hover flies: all are crucial to healthy ecosystems and to our food supply. Pollination and decomposition are just two ways in which small, seemingly insignificant creatures play an outsized role in our world.

The stories of bees, hover flies, and millipedes reveal our relationships of dependence and connectedness to other organisms. But there is something more to appreciate beyond what they do for us. I think we can appreciate the beauty of their interwoven and evolved relationships with their environment—we can even appreciate the rat-tailed maggot in that way. Aside from their value to us, their evolutionary histories and ecological interactions suggest that they have biological value in themselves. These animals are receptacles for successful adaptations made over hundreds of millions of years. They are the results of the interactions of living things and their environment since life began.

The finely tuned relationships these organisms have with their environments are changing, however, in the Anthropocene. Some relationships are broken, and animals become extinct. No longer able to rely on local pollinators, we cart honey bees around the country to support our food production. Wooden hives on flatbed trucks traverse the interstates from crop to crop through the summer. Bumble bee, millipede, and hover fly species disappear as humans alter the landscape, changing which species do the work of pollination and decomposition—if it gets done at all. Other organisms flourish in their new situations, and perhaps expand their numbers and range. Some bumble bee species decline, but others can adapt to new circumstances. Similarly, the millipede and hover fly species that live solely in old-growth forests are declining as this habitat disappears. Understanding more about how we are interconnected and how we depend on these organisms, we may appreciate them more. Knowing that these organisms reflect the long evolutionary history of life, we may value them more. In appreciating and valuing them, we may begin to understand our responsibilities even to the smaller creatures such as hover flies, which in a single lifetime may contribute to both pollination and waste removal in our Anthropocene world.

THREE

Inadvertent Domestication

The Pets We Didn't Want

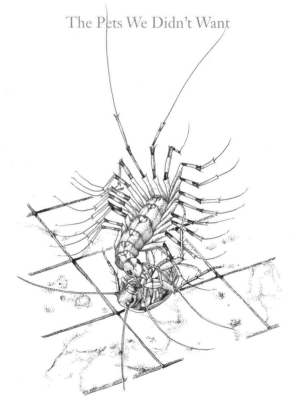

IN AN ANNUAL RITE of fall, many of us remove woolen skirts, shirts, and jackets from storage and, wearing them, for a few days thereafter carry the smell of cedar chests or mothballs with us wherever we go. We suffer these odors to protect our clothing from the depredations of the diminutive clothes moth (*Tineola bisselliella*). Putting away winter clothing with mothballs has been so common that mothballing has become another word for storage—and we even mothball huge naval ships. Clothes moth caterpillars feed in natural fabrics, especially in stained or soiled areas, creating holes in a favorite garment. But fur coats and wool sweaters have not always been available as a food source for clothes moths. Long before cardigan sweaters, clothes moths were feeding on dry animal skins. Clothes moths were one of the last steps in the decomposition of dead animals. Not many other animals can digest dry skins, fur, and feathers, but clothes moths can thrive on these low-moisture foods. Adult clothes moths meet for mating on dry skins, lay their eggs on them, and clothes moth caterpillars subsequently grow and develop there. At a time well before highway roadkill, animal remains occurred unpredictably in the environment and were widely separated. The unpredictability made it hard for clothes moths to find animal remains and thus mating sites. So until their association with humans, clothes moths had a hard time hooking up.

Once fully grown, an adult male clothes moth leaves the animal skin where he grew up, in wobbly flight. Clothes moths are not strong fliers. Because adult clothes moths don't eat, they depend on reserves they built up while feeding as a caterpillar to sustain them as they search for a mating site. These reserves may be sufficient to allow an adult clothes moth to survive for a month. A month is hopefully enough time to find a mate and complete a long search for animal remains that are suitable for the growth and development of young clothes moths. A lot can happen to a clothes moth in an extended search for a carcass. Strong winds, heavy rains, and predators could all prematurely end the search. You can see how life in and around human habitation might be preferable for a clothes moth. Night after night, the clothes moth male flutters through fields and forests searching for the right spot. With some meteorological luck and predator avoidance, the male clothes moth may finally encounter the faint scent of a long dead animal. Clothes moths rely on an excellent sense of smell to detect odors from a dry animal skin at a distance. In the search for a mating site, males take flight earlier in the evening than females, and so usually arrive at a suitable carcass first. Once the male has arrived on the animal skin, he begins to advertise his presence by producing chemical pheromones and making sounds by fanning his wings. Adding male-produced attractants to the mix of delicious odors already emanating from a carcass enhances its attraction for the female so she is more likely to land at a male-occupied site than on animal remains where males are absent. Thus, through chemical and sonic communication, male meets female on an animal skin carpet to mate. And there, on the same remains, the mated female will lay her eggs.

But that long search for scattered animal skins, their ancestral way of life, is now the exception for clothes moths. A study in Germany determined that today nearly all clothes moths live in and around human habitation rather than in the countryside. It seems that clothes moths are yet another Anthropocene species, one that we have essentially brought indoors and domesticated. The clothes moth species probably originated in North Africa. There is no record of our common indoor clothes moth in Europe until the 1800s. Some scientists speculate that clothes moths came to Europe on game trophies from Africa. Since then, clothes moths have been transported by humans all over the world, and even to locations as isolated as Australia. Their spread in association with humans is aided by an ability to manage arid environments. Clothes moths are particularly well adapted to central heating. They survive the dry indoor heating season better than most other indoor insects that feed on fabrics. The caterpillars adapt to dry conditions by spinning tiny silken tunnels or tents that help them conserve moisture.

Nowadays, clothes moths may be found in a variety of locations in and around our homes and businesses. Clothes moth caterpillars are well known for feeding in woolen carpets, furniture coverings, and clothing. But they may also be found feeding in household detritus along baseboards, in bird nests associated with human habitation, and even in beehives. The ability to feed on the proteins in dead skin, fur, and feathers is unusual. Most moth caterpillars feed on plants. Yet feeding on corpses must have been a useful capability, given the evolutionary success of clothes moths. It is thought that the earliest ancestors of clothes moths developed more than 100 million years ago. So these insects also survived the cataclysmic events that caused the great dinosaur extinction. In fact, feeding on dead animal remains would have been an excellent strategy at a time when nearly 75 percent of the earth's species became extinct. So perhaps dinosaur skins were a part of a clothes moth ancestor's diet in earlier times.[1] Clothes moth feeding habits have also proven useful to forensic scientists. Clothes moth caterpillars tend to incorporate bits of food in their cocoons. And if they are feeding on a human corpse, the cocoon may include hairs from the corpse. Scientists use these hairs to determine, through DNA analysis, that a corpse was present where the cocoon is found, even if the corpse has later been moved.

Clothes moths' relationship with humans may have started very early in our history when animal skins accumulated in caves or other areas where humans lived. This process of inadvertent domestication has been repeated with many other organisms, such as bed bugs (*Cimex lectularius*), dust mites (*Dermatophagoides* and *Euroglyphus* species), and house centipedes (*Scutigera coleoptrata*), resulting in a veritable wildlife reserve in our homes. Bed bugs followed a path to human homes similar to that of clothes moths, forming an association beginning in caves and early human habitation. They exist uncomfortably close to us and are well adapted to living in our homes. Dust mites perhaps began their association with humans a little later, taking advantage of the habitats we provide in modern homes by occupying a niche in the forest of carpet fibers in our living rooms. The ferocious-looking house centipede exploits our household habitat, finding a steady diet of small indoor inhabitants there. In a way, you might think of house centipedes as your live-in pest control operator. Our comfortable indoor environments have become their preferred habitats. These animals demonstrate the inexorable ability of nature to take advantage of and adapt to new circumstances. They represent nature that cannot be excluded from our lives, inserting itself into every new situation we create. This nature oozes, crawls, and flies its way into every nook and cranny of human structures, often finding an environment indoors that is more favorable than their ancestral outdoor settings. There is something wonderful about these adaptable creatures that enliven our homes. This domesticated nature is something that we can at least grudgingly admire if not celebrate.

Much of my career was devoted to thinking about how to restrain many of the forms of nature that squeeze their way into homes, arriving through cracks and crevices, open windows and doors, or even in bags from the grocery store. My research in graduate school, though, was on forest insects, and I envisioned myself working among maples and pines rather than apartments and gardens. But it was an opportunity for employment after graduation that led me to the peculiar ecology of non-human life in human-constructed environments—and the task of eliminating at least some of it. It turns out that the living things that have the ability to live in association with human beings are pretty intriguing. And I came to think that in some small way I was contributing to making people's lives a little better. I remember a product test in a home in Mexico. The residents participating in the trial complained of large numbers of cockroaches in their home. The test product fumigated the open kitchen and living area, blowing insecticide vapors into all of the nooks and crannies where cockroaches might hide. We returned about an hour afterward to the scene of a cockroach Armageddon. The floor, furniture, and countertops were littered with American cockroaches (*Periplaneta americana*)—a robust, dark brown cockroach up to two inches long. As my colleague described it, cockroaches were still raining from the corrugated metal ceiling. In this instance, the product improved the home environment for the people that lived there—although such broadcast pesticide use requires a significant cleanup effort, and a more targeted treatment would be preferable. It would have been difficult to live among so many cockroaches. So the residents were defending themselves against a remarkable insect that was enjoying great success in their home—a situation paralleled by a few other uninvited guests, including bed bugs.

Bed bugs have a flattened shape that allows them to squeeze into the cracks and crevices found in beds and bedrooms. They may be discovered under wallpaper, between floorboards, behind pictures, in fissures in the bedframe, or beneath piping on mattresses. Bed bugs live in groups of multiple ages and sexes. They are not active during the day, but emerge at night. A male bed bug may emerge from his hiding place in the floorboards around 1 a.m. He crawls across the floor to the leg of the bed and begins to climb. This is not his first trip to the bed. He found ample nourishment there last week. Climbing the wooden bed leg is easy for the bed bug. Using hooks at the end of his feet, he rapidly ascends using minute imperfections in the wood surface for purchase. He reaches the bedding and begins a march over its undulating surface. The bed bug travels up and down layers of sheets and blanket, growing more excited as he senses the carbon dioxide emitted by the person sleeping in the bed, and then warmth and other chemical cues as he gets closer. On reaching his host, he chooses a spot on the inside of a forearm for feeding. Bed bugs, like mosquitoes, have adapted insect mouthparts originally meant for biting and chewing to become vehicles for piercing and sucking. The two outer mouthparts act as tiny saws to create a hole for the two inner, tube-like mouthparts. One tube injects saliva into the wound to prevent coagulation and to act as an anesthetic, preventing the host from feeling the bite. The other inner tube imbibes the blood. Deep in REM sleep, the victim lies motionless while the bed bug continues to feed. The bed bug may keep feeding for up to twenty minutes, his expandable abdomen growing larger and larger. Finally, after consuming maybe five times his body weight in blood, the bed bug leaves his host and begins the trek back to his hiding place in the floorboards. The human host, waking in the morning, will not even realize she has been bitten. Studies vary, but perhaps 20 percent of people don't react to bed bug bites. If she is in the majority, though, she is likely to have a raised, itchy welt where the bed bug took its meal. But that welt won't show up for one to several days. She may be confused by the bite. Was it a spider bite? A mosquito? It is difficult to know until she actually sees a bed bug or signs of one, such as small particles of dark fecal matter or tiny blood spots on bedding. Even when the bed bug problem has been identified, it is challenging to resolve. Bed bugs are notoriously difficult to find, given their ability to hide in the tiny cavities throughout a room.

Clearly, bed bugs are well adapted to living with human beings. The first record of bed bug and human association goes back some thirty-five hundred years to a bed bug found in an Egyptian tomb. Scientists speculate, though, that bed bugs may have first connected with humans when we were living in caves. Bed bugs were probably associated with bats then, but over time developed the ability to feed on humans and followed them into their dwellings. Even though humans are their primary host today, they are still able to feed on other animals such as bats, birds, and mice. These alternative hosts enable bed bugs to survive in an abandoned home without human beings. There are other members of the bed bug family that do not feed on human blood, and these primarily feed on the blood of birds and bats. The spread of human-biting bed bugs has been documented throughout our history. The Greek philosopher Democritus recommended hanging the feet of a hare or a stag at the end of the bed to prevent bed bug bites.[2] Bed bugs were themselves thought to be medicinal, being part of remedies for a range of maladies including snakebite and malaria. Bed bugs came late to Europe, reaching Germany by the eleventh century and England by the sixteenth century. Bed bugs probably came to North America with the first European colonists—another colonial plague visited on the original Native Americans. Bed bugs had been a continuous, though unwanted, companion in our homes up until the 1940s. At that point, the use of DDT and other insecticides made bed bugs a less familiar problem. Yet bed bugs returned with a vengeance in the 1990s, when they began to turn up in high numbers all over the world. No one is quite sure why bed bugs were on the upsurge again, but the increase in global travel and bed bug resistance to insecticides seem to be important factors. As a result, we have all become a lot more familiar with bed bugs—if not directly, then through stories in the media.

Bed bugs get around either by hitching a ride on an unsuspecting person's clothing or by walking on their own six legs from one apartment to another. Bed bugs are quite capable of moving through wall voids or other connections between apartments. In New York City, bed bugs have been reported in the subway, hotels, and the Empire State Building. An entomologist reported finding bed bugs in his hotel room in San Diego at the convention of the Entomological Society of America. Another entomologist friend found bedbugs in his suburban home, and even this expert found it very difficult to eliminate them. He thought that a house sitter must have inadvertently brought the bed bugs into his home. When it comes to bed bugs, I guess you can't trust anyone. They are found in cities, suburbs, and even rural locations, and as they have become a more common experience once again, it's a frightening prospect for many—perhaps because they were seemingly absent from our lives for so long.

Most bed bug communication is accomplished chemically. Bed bugs tend to congregate in their hiding places, and scientists have found that chemicals produced in the aggregations of bed bugs both attract and arrest the bugs. Bed bug reproduction is accomplished through what is aptly called traumatic insemination. Bed bug males will mount any other recently fed bed bug of the right size. This is problematic for other males and immature females because the male inseminates the female by piercing her abdomen—even though she has a perfectly good reproductive tract. Of course, other males and immature bed bugs wish to avoid this treatment and so emit an alarm pheromone to dissuade the aggressive male. Chemical detection also plays an important role in host location for bed bugs. They are attracted to several human body odors, including the carbon dioxide we transpire. Bed bug antennae are as sophisticated at detecting chemicals as the finest analytical chemistry instruments in our laboratories.

The word "bug" was first used in the seventeenth century to refer to bed bugs. Only later was it used colloquially to refer to insects generally. According to the Online Etymology Dictionary, "bug" was derived from a word that meant something frightening.[3] So it seems that humans have a long and unhappy association with bed bugs, which have adapted to humans as their primary food source. Our response to bed bugs today, after thirty years of relative absence, is one of fear. Researchers have found reactions among those who have suffered a bed bug infestation in their home to be similar to post-traumatic stress syndrome. These individuals have difficulty sleeping, suffer anxiety, and are vigilant in the extreme toward preventing a recurrence of bed bugs. The publicity accorded to the recent bed bug resurgence has also contributed to emotional distress even without the presence of bed bugs. Many people fear that they have bed bugs without evidence of the insects themselves. Why is it that bed bugs elicit a level of anxiety beyond that occasioned by mosquitoes or other biting insects? Part of it must be that they are cohabiting our living space and yet are so elusive. It may also be that they bite us when we are most vulnerable, in our sleep. Mosquitoes may bite us at night but they are more transitory. They are not full-time residents in our homes. Maybe there is also something cleaner and more acceptable about a flying rather than a crawling insect. Our emotional involvement with bed bugs raises a question about how we relate to nature. I think bed bugs most certainly are part of nature, though they primarily inhabit our homes. But this is not how we think of nature. Normally, we conceive of nature as what exists outside human-constructed environments. In the Anthropocene, though, we know that humankind affects the whole of the earth. The extent of human impact varies only in degree and manner in different environments. That bed bugs appear in our houses, apartments, and hotels rather than caves does not make them less natural. Their behavior, physiology, and appearance are likely the same today as they were when bed bugs first encountered cave-dwelling humans.

If we look beyond bed bugs' effect on human lives, there is much to admire about the physical and behavioral characteristics that have evolved to allow bed bugs to effectively exploit us as a food source. Although we can, perhaps grudgingly, appreciate bed bugs as living beings with a successful evolutionary history, of course we wish to defend ourselves against their depredations. We are involved in a parasite and host interaction. In nature, the host develops defenses against its parasites and the parasite evolves in response. In our example, humans developed chemical defenses (insecticides) and bed bugs evolved insecticide resistance in return. We continue to develop new ways to thwart bed bugs, and it is possible that our ability to defend ourselves will outstrip the bed bug's ability to respond. In that case, we may be responsible for the extinction of bed bugs that feed primarily on humans (there are other species that feed primarily on birds or bats but that may opportunistically feed on humans). Natural selection would have been unfavorable for bed bugs in this case. No one argues that human beings should not be able to defend or feed themselves. The difficulty in applying this principle is in the degree to which it is used. When we eliminate other organisms for reasons of our own convenience rather than need, we have probably gone a step too far.

House centipedes are an indoor predator that may include bed bugs in their diet. House centipedes are a long-legged centipede commonly seen in the home, particularly in bathrooms and basements. They prefer areas of higher humidity and moisture because, like millipedes, they lack the waxy layer on their skin that reduces water loss in insects, and cannot close or reduce their breathing openings, where moisture may also be lost. House centipedes are generally considered to be quite frightening in appearance. They have fifteen pairs of very long legs, the last two pairs extending well behind the body, and a pair of long antennae extending well beyond the head in front. Although the body is only an inch to an inch and a half in length, the legs make the house centipede appear to be more than two or three times larger. House centipedes are the only centipedes able to live full time in human homes. That association has allowed house centipedes to spread well beyond what might have been their natural geographic range. House centipedes are now found in North America, South America, Europe, and Asia—from as far south as the tip of Argentina to Canada. They must live full time in houses in areas where they wouldn't survive the winter climate. House centipedes originated in the Mediterranean region, evolving in a milder climate, before we transported them around the world.[4]

Centipedes are predators. They mainly feed on insects and spiders, but certain tropical centipedes may be large enough to kill and eat small amphibians, birds, and mammals. Their insect prey includes flies, moths, cockroaches, and silverfish. House centipedes have even been known to kill and eat bees and wasps. The house centipede will stun the wasp with a poisonous bite and then retreat to a safe distance until the wasp dies and the centipede can return to feed on it. House centipedes may sense prey through their long antennae, detecting both prey chemicals and by touch. Whereas most centipedes are adapted to living in narrow spaces in crevices and beneath stones, relying on random encounters with their prey, the house centipede is fast and pursues its prey in the open. It has been calculated that a house centipede the size of a human would run as fast as forty miles per hour. It is amazing that they can coordinate all of those long, spindly legs to become the cheetah of the household!

House centipedes have evolved multiple adaptations in their body plan that make them the fastest of the centipedes. Their bodies are built to support their long legs, which they use to produce speed. One of the problems fast-running centipedes face is the tendency of their long body to undulate while running. So house centipedes have fewer body segments than other centipedes, and some of the central, hardened plates on their backs are fused. These modifications, in combination with internal reinforcing muscles, help to keep their body straight while they run. Additionally, house centipede legs vary in length, increasing in length from the shorter legs in the front to longer legs at the rear of the centipede. The different lengths make running more efficient and prevents overstepping by the legs. The long legs of the house centipede also require more muscles within the leg than are present for other centipedes. They have thirty-four muscles extending from the body to operate each leg—adaptations that exceed those of any other multi-footed arthropod. The house centipede's body plan modifications incorporate skeletal and musculature adaptations, which are accompanied by improved vision and changes in the respiratory system, to make this centipede an effective predator in the open. The house centipede can run down and pounce on a variety of fast-moving prey. Hooks and tiny adhesive hairs on their feet help with traction and also make them good climbers. They can pursue spiders up a vertical rock wall. In my experience, the sides of a bathtub are too smooth for the house centipede to climb, and unfortunately I have often found them trapped in the tub. Aside from the inability to climb smooth, enamel walls, house centipedes are marvelously adapted predators on the small creatures living in our homes. Hosting house centipedes, in a way, is like keeping a cat around to catch mice.

All centipedes use a pair of poison claws found near the head to kill their prey. These specialized claws developed evolutionarily from legs. The house centipede's poison claws are more leg-like and less robust than those of other centipedes. But their poison claws are also more mobile—built for speed and chasing down prey. House centipedes can capture and kill faster and larger prey than many other centipedes, due to their speed and mobile poison claws. There are also rare records of house centipedes biting humans, with effects ranging from none at all to pain and swelling like a bee sting. Such bites are unusual because the house centipede's normal reaction to human presence is to try to escape. They only bite in self-defense. Additionally, their poison claws often can't break the skin.

Centipedes are the oldest venomous predators. The earliest fossil record of centipedes is from about 420 million years ago. So they were present on land not too long after their other multi-legged relatives, the millipedes. They have evolved a variety of life history strategies. Like the insect silverfish, the males deposit small packages of sperm, spermatophores, in their environment, which are then picked up by females. Some centipedes remain with their eggs and young and tend them during development.

House centipedes, though, deposit eggs singly and abandon them. Their young are born with only four pairs of legs. They add segments and pairs of legs through successive molts until they reach the final number of fifteen. Chasing around on fifteen pairs of legs, adult house centipedes may live for five or six years. These leggy house centipedes are sometimes the cause of calls to pest control companies, because their size and appearance are enough to generate a strong desire to eliminate them. But the fearsome-looking house centipede also plays an important role in the household ecosystem, preying on insects and spiders. Instead of controlling them, perhaps we should be welcome their presence!

Turning from large and fearsome, we next consider the minuscule and unseen house dust mite, another organism that enjoys the environment we provide in our homes. In a way, it's surprising that we know much about the house dust mite at all. Though it may live with us in great numbers, it is difficult to see without the aid of a microscope. Adult house dust mites are about one hundredth of an inch long. And the even smaller, immature forms of dust mites are hard to see even under a microscope. It is the effect of house dust mites on human health that brings them to our attention. Unlike bed bugs, dust mites affect us indirectly. The discovery of the relationship between house dust mites and humans began with a debilitating case of asthma, which was reported in a research paper in 1922.[5] The patient in this study had suffered asthma attacks with regularity while at home in New York, but he had no asthma attacks while serving in the army in Texas. Doctors had tested all known allergy extracts on him to no avail—he didn't react to any of them. They could not discover the source of the allergens in his home that were triggering the asthma attacks. At a loss, they asked the patient to use a vacuum to collect as much dust as possible from his home. When an extract of this dust was tested on the patient, he had an allergic reaction. The doctors were puzzled. What were the allergens in the dust, and what was their source? They found that other asthmatic patients also reacted to the house dust extract. Later, in 1928, another study provided a clue for doctors. German researchers discovered mites in mattress dust and found that the symptoms of asthma sufferers could be improved if they were in mite-free environments. Despite these leads, the house dust mite was not identified as the producer of the allergens that cause asthmatic symptoms until the 1960s.

We now know that proteins in house dust mite feces and cast skins cause allergic reactions. Dust mite feces are tiny and, when disturbed by human activity, may become airborne, finding their way into the nose and then airways. Allergen proteins are released from the fecal particles when they come into contact with the moist surface of the lining of the airways. The human body becomes sensitized, and the immune system reacts to these allergens, creating the allergic response. House dust mites are now recognized as the most important producer of indoor allergens. These allergens can be a key factor in the development and persistence of asthma. They are quite potent. As little as two millionths of a gram (or four billionths of a pound) of dust mite allergen is considered sufficient to cause sensitization. There may be as many as 130 million people globally that are affected by house dust mite allergy.

House dust mites haven't always lived in carpets and on sofas. Dust mite ancestors were probably skin parasites that fed on several different kinds of animals. The house dust mite's close relatives include parasitic mites that may be found in bird nasal passages, on bird skin or feathers, or on the skin of mammals. At some point in their evolutionary history, house dust mites made a critical switch from parasitism that involved residing on a host to a free-living life in the nests of birds or mammals. Important factors in the ability of mites to make this change were a tolerance of lower humidity, the ability to feed on skin (like clothes moths), and the fact that they didn't require a specific host. It is not difficult to imagine how house dust mites may have shifted from rodent or bird nests in or around homes to a life associated with human dwellings. House dust mites are now found in human homes around the world.

House dust mites are more prevalent in humid regions, which may explain the improvement in that first dust mite allergy sufferer when he moved to an arid part of Texas. But there is no free or standing water generally found in house dust mite habitats.

You can imagine that a dust mite located at the base of carpet fibers cannot find a drink of water anywhere. So dust mites have developed an unusual mechanism for obtaining moisture. They secrete a fluid from glands located near their legs that absorbs moisture from the air. The fluid flows into troughs on the body surface, which, after absorbing moisture, eventually empty into the oral area where the mite imbibes the liquid. House dust mites survive best at 55–73 percent relative humidity. When the humidity drops below 51 percent for an extended time period, the secreted, water-absorbing fluid crystallizes at the outlet of the gland and won't flow. As a result, the mite has nothing to drink, and it dehydrates and dies. Research shows that the house dust mite only needs a few periods of higher humidity to survive, though. In one study, dust mites could survive on two hours of 75 percent relative humidity and twenty-two very dry hours of 0 percent humidity. House dust mites cannot complete their life cycle under these conditions but do so once the humidity is increased. Thus, lowering the humidity in a house for an extended period of time is one approach used for eliminating house dust mites.

What is life like for this tiny creature that lives unnoticed with us in our homes? Adult house dust mites, like all mites, ticks, and spiders, have eight legs. They roam in areas where human or animal skin flakes are plentiful. Here, they mate, lay eggs, and their young develop into adults. It takes about a month to progress through the developmental stages to become an adult, and fertilized females may live for three to four months—resulting in a four- to five-month life span. Numbers of house dust mites can then develop rather rapidly. Populations can double in just two weeks at sixty-eight degrees Fahrenheit and 70 percent relative humidity. So a sofa cushion could become quite crowded with dust mites in a short period of time. Fortunately, dust mite populations cycle over the year, rather than constantly increasing, in temperate climates. The populations increase during summer months when humidity is higher in the home, and decrease in the winter due to the drier air produced by cooler outdoor temperatures and central heating. House dust mites are commonly found in the home in carpet, on mattresses, on sofas, and on upholstered chairs. We transport both dead and live dust mites on our clothing, so it's not surprising that house dust mites are found on automobile, train, airplane, and bus seats. A recent study even found that 77 percent of child car seats contained dust mites.[6] Dust mites, then, do not just stay at home. They are also found in churches, banks, libraries, museums, hospitals, hotels, and nursing homes. You get the picture. House dust mites are found pretty much everywhere that we are, although we wouldn't realize it but for our sneezing.

Our homes support a range of living organisms, some seen, some unseen, some that are harmful to humans, and others that have no effect on humans at all. There seems to be a general desire to have our homes to ourselves, however. We are uncomfortable sharing space with other organisms because, I think, it is in our homes where we feel most vulnerable. Will the fearsome-looking house centipede crawl over us while we sleep? It is unlikely, but bed bugs will certainly bite us while we sleep if given the opportunity. I suspect that most people would be appalled if they were to learn the extent of house dust mite colonization of their home. We want a pristine indoor environment, but that's not possible. Life, in the form of living organisms, has a way of finding its way inside and, if conditions are suitable, taking up residence. It requires a change of heart, but we may be better off trying to manage the few organisms that truly cause harm, like the clothes moths, bed bugs, and dust mites, and letting the others go about their business. As we have seen, the house centipede may even be beneficial, preying on cockroaches and bed bugs. Those organisms that find their way inside and stay are a reminder that nature is in fact inescapable. Even in the Anthropocene, in houses and buildings that are tightly sealed to the outdoor environment, clothes moths will flutter through our closets, and house centipedes will race across the bathroom floor on fifteen pairs of legs. We're not easy to get along with, so we can appreciate the persistence of these invaders who have become so well adapted to living with us. As a result, the next time the ants come visiting we might consider leaving the bug spray on the shelf.

FOUR

Anthropocene Invasions

SEVERAL YEARS AGO, a neighbor complained to me about an infestation of ladybugs in her home. You could see large crawling splotches where they massed on the white siding of her house, and she was frantically filling vacuum bags of them as they came indoors. I grew up thinking that finding a ladybug, like a four-leafed clover, was good luck. But the ladybugs in my neighbor's home weren't lucky at all—at least not to my neighbor. These ladybugs produced a musty smell and would bite if handled. The ladybugs in question were multicolored Asian lady beetles (*Harmonia axyridis*). (Since ladybugs are classified as beetles, scientists refer to them as lady beetles, rather than bugs.) How did lady beetles change from a harbinger of luck to a nuisance?

As its name implies, the multicolored Asian lady beetle is not native to the United States. It was purposefully imported into North America on multiple occasions, beginning in 1916, as a predator of aphids that were pests in agriculture. Happily for the aphids, but not so much for the farmers, none of these introductions were successful, and the multicolored Asian lady beetle didn't become established in the United States until the late 1980s, in Louisiana. In spite of the earlier concerted efforts of entomologists, the insects that became established were accidentally introduced from a freighter that docked in the Port of New Orleans. Rapidly expanding their range from there, these lady beetles are now found throughout the United States and southern Canada. These multicolored Asian lady beetles have spread in combination with another invasive species, the seven-spotted lady beetle (*Coccinella septempunctata*) from Europe, and together they have contributed to the decline of native North American lady beetles.

The nine-spotted lady beetle (*Coccinella novemnotata*), a native species, was one of the lady beetles that signified good luck. In the 1950s, it was the second most common lady beetle found in the state of New York. In fact, the nine-spotted lady beetle was chosen to be the state insect of New York in 1989. At that time, although already in decline, it was still one of the more common lady beetles found in the northeastern United States. Since then, though, it has become exceedingly rare: from 1992 through 2005, no one was even able to find a nine-spotted lady beetle. Finally, a single specimen was found on Long Island in 2006, and again twenty were found there five years later. That's it. The New York state insect has become extremely uncommon. Today, Cornell University sponsors the Lost Ladybug Project to enlist citizens in the search for the nine-spotted and other native lady beetles that are going missing.[1]

Invasive species such as the multicolored Asian lady beetle often expand their populations rapidly when introduced into new areas. This growth in numbers is largely the result of leaving behind the predators, parasites, and diseases that kept them in check in their native region. My neighbor was personally experiencing an influx of the outbreak population in her home, as the mass of multicolored Asian lady beetles covering the side of her house found their way beneath the siding and inside through cracks and crevices. Rapidly growing outbreak populations of these beetles, freed of constraints, were evident in the New York state surveys that found the now rare nine-spotted lady beetles. Multicolored Asian lady beetles were already the most common species in the 2000–2001 state survey, and by the survey of 2008–2012, they made up nearly 60 percent of the lady beetle species collected. Invasive species may cause native species to decline by acting as predators or competitors, or by bringing in new diseases to which native species have no resistance. In the case of lady beetles, both predation and disease carried by the multicolored Asian lady beetles were implicated in the loss of native species.

The geographic movement of species is not a new phenomenon. What has changed in the Anthropocene is the accelerated rate of invasions. Humans are in constant motion around the globe today, and the shipping of goods via plane, boat, and train compounds the problem. Despite our efforts to keep them out, small hitchhiking organisms like insects may escape detection and become established in a new territory. The invaders arrive hidden in crates, in imported plants, and as stowaways on ships and planes. So not only do human beings physically change the environment directly, but we also inadvertently change our ecosystems through the introduction of new species like the multicolored Asian lady beetles, Asian tiger mosquito (*Aedes albopictus*), garlic mustard (*Alliaria petiolata*), and Canada thistle (*Cirsium arvense*). The Asian tiger mosquito was an accidental introduction that is a capable transmitter of dengue fever and chikungunya fever, which is now being reported from Caribbean islands. The creeping thistle or Canada thistle, also known as the lettuce from hell thistle, was introduced accidentally, probably in crop seed, and is now broadly established in North America. It is considered a noxious weed that is aggressive, spiny, and quickly takes over garden and pasture areas. Other species, such as garlic mustard, were introduced intentionally. Those who brought it to North America thought that garlic mustard would remain a quiet cooking herb in our gardens, rather than becoming a scourge in thirty-six states. It is a threat to native plants, which it outcompetes in forested areas. Invasive species are disruptive to native local ecologies and are increasingly common in the Anthropocene epoch.

Farmer periodicals in the 1800s called the Canada thistle a nuisance, a pernicious pest, and a noxious weed. The *Prairie Farmer*, in 1865, said the Canada thistle was an enemy of the human race, and the *Indiana Farmer's Guide*, in 1921, called it the worst weed of all. In 1847, the *Prairie Farmer* said the Canada thistle "should always be met with determined and continued resistance; and nothing but utter extirpation should ever satisfy the farmer."[2] What do we know about the plant that inspired such hyperbole?

The Canada thistle is not Canadian at all. Instead, it is an invader in North America, originating from the southern Mediterranean. It is considered native to Eurasia, found throughout Europe, and West and Central Asia. The Canada thistle has spread from its native range to South Africa, Australia, New Zealand, southern South America, and North America. It was probably introduced to North America on several occasions. French settlers brought the thistle to Canada in the early seventeenth century, accidentally mixed with crop seed. British and Dutch settlers likely introduced the plant to New England. Mennonite settlers also brought contaminated crop seed to the North American prairies in the late nineteenth century. Expanding its range from the original European introductions, the Canada thistle continued its spread across North America in packing material, in straw and thistle down in mattresses, and again as seed contaminant. The Canada thistle completed its journey across North America, reaching the Pacific Coast in 1899. Today, the plant is listed as a noxious weed in forty-three states and six Canadian provinces—a testament to its colonization of North America and infamy. A noxious weed is a plant considered by the authorities to be injurious. No other plant comes close to this level of concern in North America. The closest, field bindweed, scores a listing as noxious weed in thirty-five states and five provinces.

The Canada thistle is so reviled mainly for the yield reductions it causes in agriculture throughout its range. It produces losses in a wide variety of crops, including wheat, barley, rapeseed, corn, soybeans, oats, alfalfa, and sugar beets. The spiny leaves and shoots of the thistle may deter grazing and thus degrade cattle pastures. The Canada thistle is truly a plant of the Anthropocene. Humans increased its range by carrying it with us around the world. And it prospers in the disturbed habitats that we created through agricultural development and urbanization—it is commonly found on roadsides and in abandoned fields. Recent research also suggests that Canada thistle may have benefitted from the human-caused increase in carbon dioxide in the atmosphere—the main culprit in climate change. In laboratory tests, its growth increased significantly more than other invasive plants when exposed to elevated levels of carbon dioxide. Moreover, herbicides were not as effective in controlling the Canada thistle grown at higher carbon dioxide levels. In the future, the drier and warmer climate produced in many areas by climate change may also encourage broader Canada thistle distribution. Is climate change creating a monster plant?

Even without the stimulating effects of carbon dioxide, the Canada thistle is a formidable plant competitor. It is a perennial plant, returning year after year, that has an extensive spreading root system. The roots of the Canada thistle may extend laterally more than fifteen feet, and the taproot can descend to a depth of more than fifteen feet. The thistle produces shoots along its lateral roots that also grow into tall plants, forming a clump of Canada thistle. The productivity and size of the root system are part of why it is so difficult to control or remove Canada thistle.

When roots or shoots are cut by hoeing or plowing, if left in the soil, even one-inch-long pieces will regenerate the plant. The root system also outcompetes other, more shallowly rooted plants for nutrient and water resources. The plant itself may grow up to six feet in height, so it also competes by shading shorter plants. Through its root system, and growth of clumps of tall plants, the Canada thistle will come to dominate an area. The Canada thistle is not just an agricultural problem. As you might imagine, it is also a concern for conservationists. It can replace many native plants, particularly in non-forested habitats. The Canada thistle became established in Yellowstone National Park after a major fire in 1988 and continued to increase for several years. In this instance though, the forests reestablished themselves, and over time the Canada thistle began to disappear. The thistle's presence is now much more limited in Yellowstone as native vegetation reasserts itself.

The Canada thistle is also something of a chemical factory. It produces chemicals called phenolics that may inhibit the growth of, or even kill, surrounding plants. There are, however, a few plants, like cucumber, that can tolerate these chemicals. The Canada thistle also produces more benign chemicals. It has a fragrant flower that attracts many pollinators and is a very good source of nectar and pollen for a variety of bee species, as well as our friends the hover flies. The honey bee seems to be its most frequent pollinator. Pollinators are critical to Canada thistle because each plant produces either male or female flowers—not both. Pollen must therefore be transferred from the male plant to a female plant. The distance between male and female plants and the distance a single pollinator is willing to travel determines whether fertilization and subsequently seed production will occur. On a positive note, some scientists suggest that Canada thistle can help foster honey production or be used along farm field edges to support pollinators.

The fragrance of Canada thistle flowers may also work to the plant's detriment, attracting insects that feed on the flowers themselves. Both beetles and grasshoppers are attracted to, and feed on, thistle flower heads. In fact, we humans have actually employed other insects to try to control the Canada thistle. Seed-feeding and stem-mining weevils are examples of the biological control agents used. Canada thistle is a prolific seed producer, generating as many as fifty-three hundred seeds per shoot. Killing seeds is important if you want to stop the increase of the thistle, since wind-borne seeds migrate to distant fields. The seeds are beneficial, though, to some wildlife and are a favorite with finches and other seed-feeding birds. A rust fungus disease that kills roots and shoots has also been employed against Canada thistle. Unfortunately for growers, these biological control agents seem to be only marginally effective. Conservationists have found, however, that regenerating grasslands with specific plant species helps prevent the incursion of Canada thistle. For example, when common yarrow, black-eyed Susan, prairie coneflower, and Lewis flax were added to a seed mix to regenerate grassland, these plants seemed to prevent the establishment of Canada thistle. Most control efforts, though, are either chemical, using herbicides, or mechanical, involving the physical destruction of plants. Attempts to eradicate the plant have a long history. *The Plough Boy*, a nineteenth-century periodical, related in June 1819 that General Armstrong of Red Hook, New York, successfully killed Canada thistle by pouring fish, beef, and pork pickle on it—desperate measures indeed.[3]

The Canada thistle has a decidedly negative effect on native plants and agriculture: yield reductions of up to 80 percent in corn have been recorded in fields where Canada thistle is present in large numbers. It is understandable that a grower would take action to eliminate the thistle. Similarly, the conservationist is faced with the question of whether and how to act to save native plants. Canada thistle is a well-established invasive plant that has been present in North America for nearly four hundred years. Is there a point at which we should simply allow nature to take its course? As facilitators of the invasion, do humans have certain responsibilities toward the native species that are harmed by the Canada thistle? The plant does have some benefits. Its deep and strong rooting structure breaks up compacted soils, making them better suited for other plants. It is an excellent resource for pollinating insects and seed-feeding birds. I also must admit that I am partial to the Canada thistle's pinkish and purplish flowers, and find it attractive in dried floral arrangements. We can also admire the competitive power of this plant and its successful adaptation to the disturbed habitats created by humans. The Canada thistle is not wholly a pariah, as portrayed in nineteenth-century farm journals. It's more complicated than that.

The province of Ravenna, Italy, lies on the Adriatic Sea, located toward the top of the boot-shaped landmass that forms the Italian peninsula. In 2014, the province of Ravenna was said to have the best quality of life in all of Italy. Seven years earlier, on June 21, 2007, a visitor from Kerala, India, came to stay with family in Ravenna and enjoy the quality of life there. Unfortunately, he became ill just two days later, experiencing high fever and joint pain, and by July 4, one of his relatives had developed similar symptoms. In the following months, health authorities in the province began to notice a pattern of unexplained cases of high fever, joint pain, headache, and skin rash. A survey of patient symptoms led to the suspicion that the disease was chikungunya virus—a tropical disease that had not been previously seen in Europe. The name chikungunya originates from Africa, where it means being bent over, referring to the joint pain that is symptomatic of this disease. Chikungunya is transmitted only by certain mosquito species, one of which, the Asian tiger mosquito, had become well established in Italy after its initial introduction in the early 1990s. With the necessary mosquito species in place, all that was needed to start an epidemic was a source of the virus, and the visitor from India fulfilled that requirement. There were 337 suspected cases of chikungunya in and around Ravenna province in the 2007 epidemic, of which 217 were confirmed to be the disease. Most cases were mild, and the only fatality was an eighty-three-year-old man. The epidemic subsided in September of that year after mosquito control procedures were put in place. These included spraying insecticides from trucks and backpack sprayers and going house-to-house to eliminate containers, like flower pots and pet bowls, that served as prime locations for mosquito breeding.

Chikungunya virus occurs sporadically in the tropics, but there have been several large epidemics since 2005. On La Reunion Island in the Indian Ocean, in 2005, 312,500 people out of a total population of 757,000 were sickened with the virus. In 2006, an epidemic in India affected 1.4 million people. The disease has subsequently recurred in Europe in Croatia and France. In 2013, chikungunya virus was found in the Caribbean for the first time and outbreaks have since occurred on multiple islands. The first local transmission of chikungunya in the United States took place in Miami in 2014. The spread of chikungunya corresponds to the spread of the Asian tiger mosquito. The Asian tiger mosquito is rated one of the 100 worst invasive species according to the Invasive Species Specialist Group.[4] The chikungunya virus has actually evolved to become more efficiently transmitted by the Asian tiger mosquito—an adaptation noticed first by scientists in 2006 following the La Reunion epidemic. It's easy to see why: it is a beneficial association for the virus, taking advantage of the human dispersal of the Asian tiger mosquito far from its original range.

The Asian tiger mosquito has been called not only one of the worst invasive species, but also the most invasive mosquito on earth. Native to Southeast Asia, it has a black body with white stripes that leads to the name tiger mosquito. The mosquito has spread from Southeast Asia to Africa, the Caribbean, the Middle East, Europe, and North and South America. It was first found in the United States in Houston in a shipment of tires from Japan in 1985. Those tires were subsequently shipped across the southeastern United States, and the Asian tiger mosquito rapidly established itself across the region. It was another shipment of tires, this time from the U.S. state of Georgia, that brought the Asian tiger mosquito to Italy in 1990. Why tires? The Asian tiger mosquito is a container breeder. The female tiger mosquito will lay her eggs just above the water line in any small accumulation of water. In more natural settings, this may be in tree holes, for example. Around human settlements, eggs may be laid in any variety of small containers, including the aforementioned flower pots, bird baths, tin cans, and old tires. A recent study in New Jersey found that the Asian tiger mosquito was laying its eggs in the downspouts of rain gutters. But used tires, as well as plant containers, serve as an excellent vehicle for long-distance transport of the Asian tiger mosquito. Used tires are shipped all over the world, and a small amount of rainwater in a tire is sufficient for the development of Asian tiger mosquito young. Mosquito young, or larvae, are aquatic, living and feeding in water until they develop through a quiescent pupal stage, which also lives in water, into adults. The amount and kind of water necessary for mosquito breeding varies by species. The Asian tiger mosquito and a close relative, the yellow fever mosquito (*Aedes aegypti*), are adept at using the shallow water containers found around human habitation for breeding.

The yellow fever mosquito and the Asian tiger mosquito are both highly invasive species. Not only are they both container breeders, but they also transmit similar diseases, like chikungunya and dengue fever. In the summer of 2016, the yellow fever mosquito was responsible for the frightening outbreak of Zika virus in North and South America. Both species also bite their hosts during the daytime rather than the evening or nighttime as is common with other mosquitoes. These two species evolved in different geographic ranges—the Asian tiger mosquito in Southeast Asia, and the yellow fever mosquito in Africa. They are now both widely distributed around the world as a result of human activities, frequently occurring in the same locations. To date, Asian tiger mosquitoes have generally been replacing yellow fever mosquitoes where they occur together. Both mate at a similar time and in a similar fashion, and this mating behavior may have something to do with the Asian tiger mosquito's success versus the yellow fever mosquito. Researchers have found that when tiger mosquito males mate with yellow fever mosquito females, the female can neither produce offspring nor mate again.

This is not the case when the reverse occurs, and yellow fever mosquito males mate with Asian tiger mosquito females. The Asian tiger mosquito females can mate again with a male of their own species and lay viable eggs. But it appears that yellow fever mosquitoes are evolving in response. Now female yellow fever mosquitoes, in areas where the tiger mosquito is present, have changed their behavior and are much less likely to mate with Asian tiger mosquito males. Yellow fever mosquito males also seem to be getting better at recognizing their own kind. Yellow fever males, from areas where both yellow fever and Asian tiger mosquitoes occur, are now much less likely to mate with Asian tiger mosquito females.[5] There may be other factors involved in the competition between these two species, so only time will tell whether the Asian tiger mosquito will continue to displace the yellow fever mosquito.

Ecological flexibility is another key to Asian tiger mosquito success in areas where it has been introduced. The tiger mosquito can be found in both forested and open habitats, and is associated with humans in urban, suburban, and rural settings. Most other tree hole mosquitoes, originating in forested areas, have been unable to adapt as well to human associated habitats. That flexibility is further demonstrated in its adaptation to both temperate and tropical climates. Scientists have found that the Asian tiger mosquito demonstrates one of the most rapid evolutionary responses to different day lengths ever recorded. As we now know, Asian tiger mosquitoes were introduced into the

United States from Japan. Asian tiger mosquitoes from temperate climates, such as Japan, spend the winter months in the egg stage. The problem for transplanted Asian tiger mosquitoes in the United States was determining when to lay eggs so that they could survive the winter. That timing is based on day length, and Asian tiger mosquitoes have been able to vary their response to day length to deposit overwintering eggs at the right time, allowing them to survive as far north as Pennsylvania and southern New York state. Climate change projections suggest that the Asian tiger mosquito will be able to spread even farther in the future, both in the United States and in Europe, bringing the potential for transmission of previously tropical viruses further north.

The spread of the Asian tiger mosquito is significant, partly because it is such an aggressive biter, but also because of the diseases it can transmit. A recent scientific study suggested that the Asian tiger mosquito might contribute to childhood obesity, because its aggressive daytime biting can prevent children from going outside.[6] Its contribution to the obesity epidemic may be questionable, but the Asian tiger mosquito can certainly make outdoor activities unpleasant. The potential of the Asian tiger mosquito to transmit tropical viral diseases to new areas is probably the biggest concern at present. Besides the diseases it is known to transmit, dengue fever and chikungunya, the Asian tiger mosquito also appears to have the potential to transmit a range of other viral diseases, including yellow fever, Rift Valley fever, West Nile virus, Zika virus, and Japanese encephalitis. One of the requirements for transmitting some of these diseases is the ability to feed on alternate hosts. The Asian tiger mosquito has been known to attack birds, reptiles, and amphibians, but it primarily feeds on humans and other mammals. It demonstrates a marked preference for humans in many of its habitats. Further, the tiger mosquito may take multiple blood meals before laying its eggs, increasing the number of opportunities to transmit disease. Many other mosquitoes take only a single blood meal before laying their eggs. The ecological flexibility, multiple blood feeding, broad host range, and broad geographic distribution all contribute to the importance of the Asian tiger mosquito as a human disease vector.

Invasive species that affect human beings may be seen as a business opportunity for companies that sell pesticide products. Advertising is not required as the press fans the flames of fear about strange new invaders. Asian tiger mosquitoes are mean and hungry. They're a menace. Companies rush testing that demonstrates that their products kill or repel the new threat and petition regulatory agencies to add the organism to the list of pest targets on their label. A charitable view of this frenzied activity is that companies wish to educate consumers on the options available to protect themselves. This is how I tried to see these events at my company. A more cynical view sees them as a product of greed and opportunism. Product managers are judged not by how many people are helped, but on how much profit the company makes. There is often tension between a marketing department that pushes aggressive claims about product performance and the scientific community in the organization. There is a constant drive in marketing to find something new to say about a product. It kills faster, lasts longer, or repels the new invader. Thankfully, regulatory authorities, like the Environmental Protection Agency (EPA) in the United States, act as a brake on this process. New claims must pass regulatory scrutiny before being approved. I believe that I worked for one of the more ethical companies in this area of pesticide products and claims. But, still, even there the process was wearing and shakes the belief that it's an enterprise oriented to helping people. Over a twenty-year career, I'm afraid my attitude shifted from charitable toward cynicism. The best advice about invasive organisms will come from disinterested parties such as the Centers for Disease Control and the National Institutes of Health in the United States.

I have occasionally been asked by friends, who are only half joking, about the value of mosquitoes. Would they be missed in nature if they were gone? Apparently others get this question too, as there are ongoing arguments in the scientific community about how important mosquitoes are in various ecosystems. Are they an important food source for birds or bats, for example? They are, in fact, a food source for a variety of animals as adults in the air and in their immature stages in the water, but are not likely crucial to the survival of any other animal. They are also pollinators, visiting flowers for nectar, but again are probably not crucial to the reproduction of any specific plants. Science writer David Quammen has suggested that mosquitoes may even help preserve ecosystems by making them unpleasant for human beings![7] To my way of thinking, the function it performs in an ecosystem is not the only measure of an organism's value. In addition to its role in the ecosystem, the evolutionary history of an organism, along with the fact that it has been successful enough to be present today, is another important sort of value that organisms convey. As we have seen, the Asian tiger mosquito is a highly successful animal. It is readily adaptable, evolving rapidly to extend its geographic range. That adaptability enables it to take advantage of the human-altered habitats and biological communities of the Anthropocene. But its success endangers human beings, so we are faced with a stark choice between human health and allowing the survival of an amazing animal. Of course, our ability to control the Asian tiger mosquito is limited. We are not faced with a moral dilemma between human health and exterminating the Asian tiger mosquito—because extermination cannot be achieved. But that possibility may not be so far away, as scientists explore genetically modified mosquitoes in an attempt to eliminate the ones that carry malaria. The decision may not be a difficult one, but I hope we can make it with an appropriate respect for what the organism represents in terms of evolutionary success.

In the northeastern and midwestern United States, people are getting together for an activity variously called a "pullathon," a garlic mustard pull, or a garlic mustard challenge. Garlic mustard is an invasive plant species that may play a role in reducing plant community diversity in forest ecosystems. I say *may play a role* because some research suggests that a one-time open-and-shut case against garlic mustard may be a little more complicated than originally thought.

Early settlers from Europe likely brought garlic mustard to North America. Garlic mustard was a medicinal and cooking herb in its home range in Europe, Asia, and North Africa. It is one of the oldest cooking spices, dating back to over six thousand years ago. Garlic mustard was first reported growing in the wild in North America on Long Island in 1868. Today, garlic mustard is present in the northern tier of the United States, and southern Canada. Garlic mustard plants live for two years. During the first year, the germinating seed produces a low-growing plant, leafing out in a rosette shape. In the second year, the adult plant develops a flowering stem, about three to four feet in height, producing clusters of small white flowers. The flowers are most often self-pollinating, but hover flies and bees also visit the flowers, contributing to some cross-pollination. The garlic mustard plant dies after producing seeds. Garlic mustard is quite prolific, producing as many as five thousand seeds per plant, and can grow both in disturbed, open areas and in forests, where it is shade tolerant.

Garlic mustard is another plant that produces a variety of chemicals that are used both competitively and in self-defense. It even manufactures cyanide-containing compounds, particularly in the leaves of the younger, rosette stage. The range of chemicals is unique compared with those produced by other related species that are native to North America. So garlic mustard chemicals are new to the native plants, plant feeders, and other organisms there. These chemicals may directly inhibit the growth of their plant neighbors. They may also affect the soil microorganisms that support the growth of other plants. Many woodland plant species have mutually beneficial relationships with soil fungi that grow in and on their roots. The fungus assists the plant in the uptake of water and nutrients, and the plant provides carbohydrates produced by photosynthesis to the fungus. Garlic mustard is known to suppress these beneficial fungi in North America, but not in Europe, where the fungi may have developed resistance to garlic mustard chemicals. The chemical cocktail found in garlic mustard is also distasteful to many plant-feeding animals, and thus protects it from the likes of insects, slugs, and deer.

The spectacular success of garlic mustard across North American woodlands appears to have come at the expense of native plants and resulted in reduced plant diversity in the forest understory. A variety of factors contributes to the competitive advantage garlic mustard has over other plant species and its resulting geographic spread. Like other invasive species transported by humans, garlic mustard has left behind the plant feeders and diseases that kept it in check in its native range. As we have seen, the chemicals produced by garlic mustard contribute to a competitive advantage versus other plants by inhibiting their growth. Significantly, the chemicals also deter feeding on garlic mustard by deer. And when deer preferentially feed on native plants, it creates openings for the further spread of garlic mustard. Garlic mustard germinates in the early spring, before most native plants, and is able to take advantage of a more open canopy in deciduous forests before the trees produce leaves, and therefore gain early access to soil nutrients. In its second year, the flower stem also emerges in the early spring and grows rapidly, shading and perhaps suppressing native plants that emerge later. These factors combined with prodigious seed production make garlic mustard a challenging competitor for native plant species. As a result, it is banned, prohibited, or designated as a noxious weed wherever it is found in North America.

Other studies, however, provide a more nuanced view of the effects of garlic mustard. Surprisingly, researchers in several locations have found that garlic mustard has little or no effect on plant diversity. In some of these studies, it appears that other factors, such as deer browsing and fire suppression, have had a more significant effect on native plant diversity. In these instances, the spread of garlic mustard may be enabled by these other factors. Deer browsing, which has grown appreciably in many areas as deer predators have declined, seems to be a particularly important factor. Research has also shown that the effect of garlic mustard may be negligible on some native species and more significant on others. Other work has suggested that garlic mustard produces reduced levels of competitive chemicals after becoming established in an area. It is possible that native organisms become more tolerant of the chemicals and that, once established, the chemicals provide less benefit to the garlic mustard plant. It is risky to generalize findings from research in local areas, but we can certainly say that the effect of garlic mustard on woodland plant communities is more complicated than we may have originally thought. Further, the relationship between native plant communities and garlic mustard is dynamic and will change over time as native organisms and communities evolve in response to garlic mustard.

The human shaping of the earth in the Anthropocene takes many forms. We reshape the geosphere in surface mining and in building places to live and work. We change the atmosphere through the emission of gases that are a byproduct of human activity. We change the biosphere directly as we fish, cultivate crops, and harvest timber, and indirectly in many activities that affect nature. Moving organisms to new geographic territory, whether purposefully or inadvertently, is yet another way of changing the biosphere. These shifts in location free the organisms from natural predators and diseases that held their population in balance in their native range. An organism may then become a dominant life form in its new locale and suppress native species, reducing biodiversity. The invasive species may also interfere with human pursuits, becoming a nuisance, affecting food supply, or harming humans directly through disease. We found in the case of garlic mustard that because the effects of invasive species may vary by locality and by affected native species, and may evolve over time, it is sometimes difficult to predict long-term effects. Even if we could expect a longer-term rebalancing of plant and animal communities, it may be too late for several of the native lady beetle species affected by the multicolored Asian lady beetle. It seems likely that we have suffered a permanent reduction in lady beetle diversity in North America. If so, the richness of our lives as members of the biotic community is also diminished. We need not be directly affected in our daily lives for a loss of biological richness to matter. It is enough just to know that we may no longer be able to see the nine-spotted lady beetle for us to feel that loss.

Other invasive species affect us more directly. Canada thistle, for example, interferes with food production, and the Asian tiger mosquito participates in expanding human exposure to diseases in new locations. Media reports tend to portray invasive species as something evil, but to simply revile them is an oversimplification. They found their way here through human intervention, and they are doing what comes naturally. There is always something to admire about these organisms. If we must remove them to protect ourselves or to protect native biodiversity, at least we can accord them the respect they are due as highly successful organisms that have the ecological adaptability to establish and flourish in entirely new locations. Whatever actions we must take, I believe it's important to revisit how we think about the organisms we are affecting. To some this may sound like a romantic notion, but many indigenous peoples practice respect for all of nature. The Haudenosaunee, or Iroquois people, delivered a message to the Western world through the United Nations in 1977, stating: "In the beginning, we were told that the human beings who walk about the Earth have been provided with all the things necessary for life. We were instructed to carry a love for one another, and to show a great respect for all the beings of this Earth. We are shown that our life exists with the tree life, that our well-being depends on the well-being of the vegetable life, that we are close relatives of the four-legged beings. In our ways, spiritual consciousness is the highest form of politics."[8] A spiritual relationship with nature is a path toward respecting all of the plants and animals in our environment, whether large or small, invasive or native. For the Haudenosaunee, that spiritual relationship reflects our interdependence and relatedness to other species, and it forms a solid foundation for an ethic related to all of nature.

The Unlucky

Anthropocene Extinctions

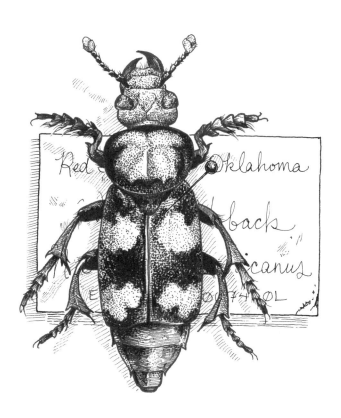

IT IS OFTEN DIFFICULT for organisms to adapt to Anthropocene alterations of the biosphere and geosphere. Human-caused changes in the environment occur relatively quickly in terms of geologic time, and some species cannot evolve responses quickly enough to survive. The effect has been a drastic increase in the rate of species extinctions around the world. As a result, we are thought to be in the beginnings of what is referred to as the sixth great extinction. On five separate occasions previously there have been sweeping losses of species on the earth. In the fifth great extinction, nearly 75 percent of all species disappeared, including the dinosaurs. The third major extinction event nearly eliminated all life, with 96 percent of the extant species vanishing. These extinction events were caused by geologic activity, such as volcanic eruptions, climate change, and even the impact of an asteroid. Today, however, in the Anthropocene the changes driving extinctions are caused by the activities of human beings. The present rate of species extinctions is thousands of times greater than what we have observed over the past 60 million years. Insects have been particularly affected in the Anthropocene due to pesticides, habitat loss, and climate change. The precipitous drop in insect numbers has been observed around the world and is now popularly referred to as the insect apocalypse. We are aware of these difficulties, and we have responded with efforts to mitigate them—though much more needs to be done. The United States government, for example, has responded to the potential loss of organisms with the Endangered Species Act.

But encouraging the flourishing of other organisms is not always as simple as it might seem. Take the case of the large blue butterfly (*Maculinea arion*) in Britain, an attractive insect, and one we would hope would be around for a long time. As a result of the interest they generate, large blue butterflies have been called the "flagship species of European conservation." Large blue butterflies are beautiful insects, whose wings are almost azure in color. These butterflies were never abundant, but their populations declined steeply from the 1950s through the 1970s until finally, by 1979, the large blue went extinct in Britain. The story of the large blue butterfly is as fascinating as it is complex. Unfortunately, a lack of understanding of this complexity prevented us from saving the large blue in Britain. The story reveals how changes in human practices may trigger a sort of ecological chain reaction. In this case, it was agricultural practices leading to a change in grass height in large blue butterfly habitat that led to the butterfly's demise. At the time, however, we didn't have sufficient knowledge of the large blue butterfly to know how it would be affected by these changes.

The large blue butterfly lays its eggs on wild thyme in Britain. After the eggs hatch, the large blue butterfly caterpillars feed on thyme plants through three molts. After the third molt, the caterpillar drops to the ground to await an encounter with red ants, known as *Myrmica* species to scientists. The caterpillar produces chemicals on its body that match those of the ant's brood. Believing the caterpillar to be one of its own, the ant carries the caterpillar back to its nest. Once in the nest, the ant-mimicking caterpillar employs one of two strategies, depending on the caterpillar species. Some species are fed and cared for by the ants. These caterpillars exude odors that make the ants think the caterpillar is a queen ant. In the British species of large blue butterfly, once inside the nest the caterpillar becomes a predator and simply consumes the brood in the ant nest. Not just any ant will do, however. If the chemicals on the body of the caterpillar don't quite match those of the host ant species, then the ants grow suspicious and consume the caterpillar—the predator becomes prey. The large blue butterfly caterpillar produces chemicals that most closely match a species of red ant in Britain called *Myrmica sabuleti*. The large blue butterfly caterpillars are thus more successfully accepted in *M. sabuleti* nests. Caterpillar success is much lower in nests of closely related ant species like *M. scabrinodis* that differ slightly in surface chemicals from *M. sabuleti*. *M. sabuleti* and *M. scabrinodis* are very similar ant species, and even expert entomologists have difficulty telling them apart. But their chemical differences determine whether large blue butterfly caterpillars will be predators or prey.

Both *M. sabuleti* and *M. scabrinodis* ants occur in grasslands in Britain. *M. sabuleti*, the best host for large blue caterpillars, has a habitat narrowly defined by soil temperature. Soil temperature is influenced both by shading produced by plants in a location and by exposure to the sun—such as warmer, south-facing slopes. *M. scabrinodis* prefers nesting in sites that have slightly cooler soil temperatures and can accept more shading. So *M. sabuleti* is found where grasses are one to three centimeters tall, whereas *M. scabrinodis* is found where grasses are taller than three centimeters. Thyme, the host plant for the younger stages of the large blue caterpillar, can grow among both short and taller grasses. The large blue caterpillar can start life wherever thyme is found, but it is less likely to develop into an adult butterfly unless it is in the very specific short grass habitat of the *M. sabuleti* ant.

The British extinction of the large blue butterfly in 1979 was probably related to multiple factors, but the most important were changes in livestock grazing patterns and the introduction of a devastating disease in rabbits. Changes in livestock practices, perhaps a greater use of feedlots, in Britain resulted in a significant reduction in grazing and thus, as grass was allowed to grow, there were fewer short-grass habitats. Additionally, myxomatosis, a viral disease of rabbits, spread throughout Britain in the 1950s, reducing rabbit populations by 99 percent. An enterprising farmer in Britain (perhaps multiple farmers) recognized the opportunity to reduce rabbit depredations and increase agricultural yields by importing infected rabbit corpses from Europe, where there was a myxomatosis epidemic. As might be expected, yields of a variety of crops did go up with the elimination of rabbit grazing, but unfortunately the short-grass habitat required by *M. sabuleti* ants was also considerably diminished. Fewer rabbits and less cattle grazing resulted in reductions of short grass, and thus the decline in *M. sabuleti* ants. The disappearance of *M. sabuleti* ants was directly related to the disappearance of the large blue butterfly in Britain.

Humans were unable to adequately support the flourishing of the large blue butterfly because we failed to understand the complexity of the factors required for the butterfly's success. It went extinct in Britain. We now have that knowledge, and the large blue butterfly has been reintroduced there. Conservationists recognized that the first step in considering a reintroduction of the butterfly was to recreate the short-grass habitats that would support the *M. sabuleti* ant. Once the ants were established, British conservation organizations obtained large blue butterfly eggs from Sweden. Today, the large blue butterfly is reestablished in more than thirty sites in Britain. Habitat restoration to short grass has worked for the large blue butterfly, but other animals may not need such heroic efforts. They may need only to be left alone. Many organisms threatened by Anthropocene changes fall into this category. The Hine's emerald dragonfly (*Somatochlora hineana*) is an endangered midwestern North American dragonfly. Habitat loss and even collisions with automobiles have significantly reduced their range and numbers. The Franciscan manzanita (*Arctostaphylos franciscana*), a plant specially adapted to the climate and soils of the San Francisco Bay area, has also been threatened by loss of habitat in the Anthropocene. The factors involved in the decline of the American burying beetle are less certain. Habitat loss may be a factor. Or, the loss of higher predators, like bears and wolves, may have resulted in competitive pressures unfavorable to the American burying beetle (*Nicrophorus americanus*). This is an insect of some renown, as it has been named specifically in at least two legislative initiatives in the United States Senate. Sadly, the bills were not aimed at preservation of the American burying beetle.

Our interventions in nature often have unanticipated effects. The widespread use of the insecticide DDT led to the thinning of eggshells in predatory birds such as peregrine falcons (*Falco peregrinus*), brown pelicans (*Pelecanus occidentalis*), and bald eagles (*Haliaeetus leucocephalus*)—an outcome that wasn't understood until after we had been spraying DDT for years. The effects of DDT figured prominently in the publication of the iconic book *Silent Spring* by Rachel Carson, which helped lead to the environmental movement of the 1960s. The use of household and garden insecticides, like the products I worked on, are likely not responsible for extinctions. But they do account for a lot of unnecessary killing. When insecticides are sprayed outdoors to kill wasps or mosquitoes, they also kill non-target insects, perhaps even the butterflies and honey bees we enjoy having around. Pesticides sprayed in and around homes to kill crawling insects like ants and cockroaches contain long-lasting insecticides that will continue to kill any other small organisms that may creep over the residue. Household pesticide products sold in grocery stores advertise that their residues continue to kill insects for as many as six to even twelve months. Do we really want to live with these killing fields in our homes that continue to be active for months? I once thought that these products provided a low-cost alternative to employing expensive professional pesticide services. But my experience is that most people lack the knowledge to use them effectively in a targeted way. Toward the end of my career, I advocated for the removal of these long-lasting pesticides. I believe their use, if at all warranted, is best left to professional exterminators, although they don't always get it right either. It's not just our exposure to persistent pesticide residues in and around our homes that's troublesome, but these are indiscriminate bug killers that eliminate non-harmful and beneficial organisms as well as the targeted pest. Too often, attacking one bug attacks all of these smaller living beings. In the case that an organism poses a serious problem, then I recommend the use of a product that is selective and breaks down quickly in the environment to non-toxic chemical components.

From a perch on a small twig, a dragonfly darts upward, forms a basket with its legs, and catches a mosquito on the wing. The dragonfly begins consuming the mosquito as it flies back to its perch. Dragonflies are impressive predators and, like hover flies, have flight capabilities that humans study as models for the improvement of flying machines. Dragonflies may hover, abruptly change direction, and fly at up to thirty miles per hour. Unlike most insects, which move their wings in pairs, dragonflies may move each of their four wings independently. Independent wing movement allows the aerial acrobatics for which dragonflies are known. Dark veins provide structural integrity to the thin, transparent dragonfly wings. Dragonflies are the products of 300 million years of evolution. The earliest dragonfly ancestors took flight before the time of the dinosaurs and were the largest insects ever recorded, displaying wingspans of two feet or more. But dragonflies aren't just great aerialists. The dragonfly life cycle occurs in two distinct stages. The immature nymphs live underwater, whereas adults roam the air. It is the aerial life stage with which we are most familiar. Although the underwater nymphs are important aquatic predators, they are generally unseen and therefore unknown by most people.

The Hine's emerald dragonfly is the only dragonfly classified as endangered by the United States government. Like the large blue butterfly, this dragonfly is threatened by habitat destruction. In the case of the Hine's emerald dragonfly, the quality and extent of the aquatic nymphal habitat has been diminished by human activities. Human development destroys its habitat directly through damage to wetlands, and indirectly through water pollution. The Hine's emerald dragonfly's range once extended as far east as Alabama, west to the Ozark Mountains in Missouri, and north to Ontario. Today, the Hine's emerald dragonfly is found only in very localized areas in south central Ontario, the Upper Peninsula of Michigan, three locations in Wisconsin, the Des Plaines River near Chicago, and in the Ozarks. They are no longer found in Alabama, Ohio, or Indiana. The nymphs require grassy wetlands fed by a constant flow of surface water or groundwater for their development. These very specific requirements and habitat loss result in the scattered distribution of the dragonfly. In the Des Plaines River area, a six-lane highway bridge recently bisected important habitat. Industrial and other urban development have resulted in pollution, changes in water flow, and the invasion of exotic plants—all of which contribute to further threatening the Hine's emerald dragonfly along the Des Plaines River. Other Hine's emerald locations, such as Door County in northern Wisconsin, and the Upper Peninsula of Michigan, are vacation destinations, and habitat there has been preserved to a greater degree.

Very few dragonfly species can survive in the wetlands that the Hine's emerald dragonfly prefers. Low water flow through the wetlands can dry up in the summer, killing dragonfly nymphs. Yet this dragonfly has developed a unique strategy to cope with periodic dry spells, and with the freezing winter temperatures found in Wisconsin, Michigan, and Ontario. The nymphs engage in cohabitation with crayfish! When water flows dry up in midsummer, these nymphs move to devil crayfish (*Cambarus diogenes*) burrows that remain wet. Although not optimal (because prey are less available to the nymphs and the nymphs themselves may become prey for the devil crayfish), the residual water in burrows permits the nymphs to survive dry periods. The crayfish burrows may also be important to the nymphs in winter, since water in the burrows is less likely to freeze. Hine's emerald dragonflies remain in the nymphal stage for three to five years and therefore need an effective way to survive the deep freeze of multiple winters.

Hine's emerald dragonflies emerge as aerobatic adults from their aquatic life stages in the summer months and live six to eight weeks. The adult dragonflies are handsome, with long, metallic green abdomens and yellow stripes on the sides of the thorax, or central body division. The prominent eyes are characteristically bright emerald-green. Like other dragonflies, they catch insect prey in flight. Adult males patrol territories in wetland habitat, feeding and searching for females. Females avoid the wetlands where males are numerous, and, until ready to mate, prefer drier habitats, where males will not harass them. Therefore, efforts to encourage Hine's emerald dragonfly populations must preserve not only the grassy wetlands the nymphs prefer, but also drier habitats surrounding the wetlands utilized by the adult females. Once females are ready, they mate in the air while flying with males. The females lay hundreds of fertilized eggs that are dropped to the surface of the water while flying.

Henry Ford was no friend to these animals. Automobiles pose a triple threat to the Hine's emerald dragonfly—habitat loss, pollution, and collision. Roads may interfere with wetland water flow, and generally disrupt habitat, as in the case of the highway bridge through the dragonfly's Des Plaines River habitat. Road runoff also contributes to pollution of important wetlands. The use of off-road vehicles in wetlands can directly damage the habitat. In addition to habitat destruction, a surprisingly large number of dragonflies are killed by impact with automobiles. Dragonfly mortality due to collisions with cars is estimated to be between two and thirty-five individuals per kilometer of road-way per day in the Chicago area. A study in northern Michigan revealed that cars kill eighty-eight dragonflies per kilometer of roadway per day. When you consider the number of kilometers of roadway through dragonfly habitat, and perhaps ninety days of dragonfly flight per summer, you'll realize that we're talking about a lot of dead dragonflies. This is a significant concern for an endangered species, like the Hine's emerald dragonfly, whose numbers are already limited. In Door County, Wisconsin, it has been estimated that cars kill thirty-three hundred adults in a single year. Aside from the mess that dead bugs make on our cars, it might not occur to many of us what kind of impact these collisions have on insect populations. The carnage is not limited to dragonflies. It has been estimated that cars kill five hundred thousand monarch butterflies per week on Illinois roadways. And many amphibians like frogs and salamanders are run over on roads that they attempt to cross to get to breeding grounds. What can we do to lessen the impact of cars and trucks on these animals?

A number of strategies are employed to try to reduce dragon-fly mortality. Speed reduction is being tried in Door County,

Wisconsin. Dragonflies are fast and agile fliers, able to quickly change direction, and thus a reduction in car speed may allow them to avoid fatal collisions. Another possibility is using barriers near the roadway to cause dragonflies to fly at a height above cars. Hine's emerald dragonflies tend to fly at heights of less than six feet, which puts them directly in the path of speeding cars when crossing a road. Barriers that rise above six feet encourage dragonflies to cross the road above cars.[1] This approach can work well with smaller, two-lane roads. But the dragonflies tend to dip down to their normal flying height if the road is wider. The barriers also may completely prevent dragonflies from crossing, which could result in the separation of groups on either side of the road into non-interacting populations—preventing reproduction and reducing genetic variation. Clearly, it would be best if roads didn't intersect Hine's emerald dragonfly habitat at all, but where roads do exist, lowering speeds to prevent dragonfly collisions with cars seems prudent. It's difficult to imagine travel that would be significantly impacted if five minutes of driving time were added to the trip as a result of lowered speed limits protecting endangered dragonflies. Furthermore, we can act to prevent the construction of future roads and ensuing collisions in those areas.

The conflict between Hine's emerald dragonfly needs and human desires has resulted in its endangered status. This creature has very specific habitat requirements, and the encroachment of water pollution, roads, and development significantly impacts those habitats. Suitable habitat becomes more and more limited as human development progresses. Fortunately, many people value the Hine's emerald dragonfly, and have become aware of its dire circumstances. Efforts are under way to preserve existing habitat and populations of this insect. The United States Fish and Wildlife Service developed a recovery plan to conduct research on this insect, to protect it where it is known to occur, to search for other populations, and to educate the public and land managers about it. Inclusion on the endangered species list has given the Hine's emerald dragonfly a chance to survive another 350 million years.

Only a few years ago, a species once thought extinct arose from the grave in the city of San Francisco. This botanic "Lazarus" is the Franciscan manzanita, and its story depicts common issues for many species affected by urbanization and development.

Manzanitas are a group of western plants, growing from British Columbia to Baja California. Spanish explorers in California called the plants manzanita, or little apple in Spanish, for their tiny red fruits. Manzanita species take multiple forms: as low-growing ground cover, as shrubs, or as trees. California is home to the greatest number of manzanita species, where they have adapted to a variety of unique microclimates and soil types. Manzanitas share the genus *Arctostaphylos* with bearberry, which is found in North America, Europe, and Asia. This genus falls in the Ericaceae family, which includes rhododendrons, azaleas, heather, huckleberry, and blueberry. The ancestors of today's manzanitas appeared about 15 million years ago, in the middle Miocene. It was a period of time when forests were receding and grasslands were on the increase. The drying and warming climate in California at that time resulted in rapid speciation to fill the changing geography. Manzanita species developed adaptations to drier and more fire-prone climates, becoming an important part of the chaparral community in California. Chaparral is composed of dense, shrubby thickets that are often highly flammable. One of the manzanita adaptations was the need for fire to germinate dormant seeds. Timing germination to recurrent fires ensures that seeds will develop in a landscape that has been cleared of low-growing vegetation that might otherwise impede seedling growth.

The association with unique climatic and soil conditions led to the evolution of a number of manzanita species that are quite localized and rare. It is said that every community where manzanitas grow in California has its own manzanita species. Dan Gluesenkamp, a biologist and the executive director of the California Native Plant Society, told me that "manzanitas are an absolute nexus for our ecosystem." The expansion of human activities in the nineteenth and twentieth centuries served to even further limit the distribution of these already rare species. So it was with the Franciscan manzanita. This plant is associated with the soils of serpentine rock formations—a soil that is limited in nutrients and high in chemicals known as heavy metals. Plants require special adaptations to survive on serpentine soils, and the North Coast range in California and Oregon supports rich plant communities that live solely on these soils. The Franciscan manzanita is thought to have occurred naturally on serpentine outcrops in the bluffs and hills surrounding San Francisco Bay. The plant is well adapted to the poor soils and summer coastal fogs occurring on these bluffs in the Bay area.[2]

The Franciscan manzanita is a spreading, low-growing shrub with attractive reddish twigs and branches covered by small, dark green, oblong, rounded leaves. It grows to a height of two feet and may spread over an eight-foot area. In the spring, the plant sports small clusters of creamy, pink-tinged flowers. In the summer, small reddish fruits, the little apples, form and birds and other wildlife feed on them. Native Americans once ate the fruits, although they are said to be rather bland and mealy. The fruits and leaves of the Franciscan manzanita were also thought to be medicinal.

Unfortunately, urbanization in the San Francisco Bay area led to large-scale loss of the localized plant communities on serpentine bluffs. The historic range of the Franciscan manzanita was likely limited to the San Francisco peninsula; it became exceedingly rare as the city and its suburbs grew. The Anthropocene caught up with the Franciscan manzanita, and led to its decline. Not only was the plant removed from its habitat, but also the fire regimes necessary for propagation were altered. Manzanita seeds may persist in the soil for many years waiting in vain for a fire to stimulate germination. Moreover, climate change may now be affecting the coastal summer fogs, which the Franciscan manzanita requires. Many human-driven factors are working against this attractive plant, but many are also working to save it.

The past century has seen some heroic interventions by botanists in an effort to preserve the Franciscan manzanita. During the San Francisco earthquake of 1906 and its subsequent fire, botanist Alice Eastwood entered a damaged California Academy of Sciences building to save botanical specimens ahead of the fire that destroyed the building—including the Franciscan manzanita specimens used to identify it as a species. Later, in 1947, developers destroyed the last known Franciscan manzanita plants, as a cemetery from the Gold Rush era was converted to a building site. Fortunately, botanist Lester Rowntree backed her Packard up to the cemetery in the dead of night and, with a shovel and a gunny sack, dragged back a manzanita. Dan Gluesenkamp quotes her as saying that she "garnered it ghoulishly in a gunny sack." Cuttings from this plant were placed in the San Francisco botanical garden and in the Tilden botanical garden in the East Bay above Berkeley. It is said that Alice Eastwood was moved to tears when she saw one of the replanted Franciscan manzanitas in a botanical garden. The Franciscan manzanita subsequently became available as a hybridized nursery plant for landscaping, but it was thought to be extinct in the wild. It was the end of the road for the Franciscan manzanita—or so it seemed.

Driving home from a climate change conference on October 16, 2009, Dan Gluesenkamp was scanning the roadside for invasive plants. Shortly after exiting the Golden Gate Bridge, Gluesenkamp spotted something strange in a small earth island between a Highway 1 on-ramp and the roadway, Doyle Drive. He thought it might be a Raven's manzanita, which is also endangered. The island was partly covered by wood chips from the destruction of the surrounding vegetation. That vegetation, including a Monterey cypress, tea tree bushes, English ivy, acacia, and cotoneaster, had obscured the low-growing manzanita bush. The only reason that the bush was not also reduced to chips was that a California Highway Patrol car had been parked next to it. The construction crew directed the chipper away from the police car, and thus the low-growing bush was spared.

Gluesenkamp came back a few days later to have another look and still wasn't sure of the plant's identity. He drove by a third time, and on this occasion slowed to take a picture through the passenger window. He then called Lew Stringer, an ecologist from the Presidio Trust, who was responsible for the Presidio National Park of San Francisco, a 1,500-acre park where the roadway and the manzanita were located. Stringer and Presidio biologist Mark Frey drove past the island ten minutes after receiving the message, making four passes to try to be sure of what they were seeing. They finally parked the car, hiked to the road, and hustled across the busy six lanes of traffic to get a look at the plant. Probably no one had been on this traffic island besides the chipping crew and these two biologists since the road was built in 1937. Once there, they found a healthy manzanita, growing among hubcaps and other roadway debris. Later, a third biologist, a graduate student studying manzanitas, Michael Chassé, was summoned to the site. The three biologists identified the plant as the Franciscan manzanita. Experts at San Francisco State University later confirmed the identification. The plant once thought extinct in the wild had been rediscovered—surrounded by highway ramps, in a location passed by 100,000 cars daily.

In a further twist of fate, that portion of the roadway was due for a complete reconstruction—part of a $1.1 billion stimulus project. The removal and chipping of vegetation on the traffic island was in preparation for construction. The Franciscan manzanita was located on the southern approach to the Golden Gate Bridge, a busy roadway that was to be converted to a parkway. It was phenomenally good fortune both that wood chips didn't cover the manzanita and that Gluesenkamp happened to drive by scanning the roadside for invasive plants.

Before beginning the construction project, biologists, government authorities, and the project team needed to determine what should be done for the last wild Franciscan manzanita. If they left it in place, it would be subjected to the surrounding construction project, and possibly damaged or destroyed. Additionally, it would still be on a heavily traveled roadway, and exposed to all the debris and exhaust from passing cars. If they moved the plant to a botanical garden, it would no longer be the last wild Franciscan manzanita. It was agreed to move the manzanita to an undisclosed, more isolated habitat in the Presidio. Additionally, as a part of the conservation plan, cuttings and seeds from the plant were provided to local botanical gardens and nurseries in an attempt to further propagate the species, and to provide a backup plan in case the mother plant did not survive the move. Thankfully, the transportation project had a budget to cover environmental surprises, although no one anticipated finding a once extinct plant in the roadway.

Moving the last wild Franciscan manzanita was a significant challenge. It was a large, spreading plant, and with its root ball and attached soil weighed about ten tons. Dan Gluesenkamp says that the plant had been growing there since the Golden Gate Bridge was built and maybe before that. Work began at 2 a.m. on a drizzly San Francisco morning. The crew cut a block around the plant and drove steel pipes beneath it. A crane was needed to lift the plant onto a flatbed truck, which carried the manzanita through closed city streets and then into the park on muddy trails to its new home. Once it reached the selected location, a second crane deposited the Franciscan manzanita into its new environs on suitable serpentine soils in a remote location in the Presidio.

In 2012, the U.S. Fish and Wildlife Service designated the

Franciscan manzanita an endangered species. As part of the recovery process for the species, a dozen sites in the city of San Francisco were designated critical habitat areas, in which Franciscan manzanita will be reestablished from cuttings. These sites are all either in the Presidio area or in city parks. This is not a circumstance we might ordinarily imagine for an endangered species. The only members of the species live not in a natural setting far away, but in the city itself, surrounded by millions of people and their structures.

The Franciscan manzanita experience is unusual in a number of respects. First, the Franciscan manzanita was thought to be extinct for sixty years—long before the establishment of the Endangered Species Act. The Fish and Wildlife Service generally declares a species endangered when it is recognized as declining. In this case, the ruling was based on the last single, wild plant. This species requires not just protection, as is the case with many other endangered species, but active reestablishment. And the entirety of the reestablishment is taking place in one of the most densely urbanized locations in the United States. Many of the objections raised to the establishment of critical habitat areas for the plant reflected concerns about continued recreational uses of the selected park areas. There is certainly potential for trampling, as well as for vandalism in these urban park areas. But many other challenges exist for reestablishment of the Franciscan manzanita in its newly reacquired habitats. It is thought that fire was important to the germination of Franciscan manzanita seeds, and scientists have now found that the seeds germinate better when exposed to smoke-infused water. Will a fire regime become a part of life in these San Francisco parks?

The last wild Franciscan manzanita has been rescued from the destructive forces of the Anthropocene—at least temporarily. Many concerns remain, including disease, direct human depredations, loss of pollinators, lack of genetic diversity, loss of natural fire regime, and climate change. Climate change may result in modifications to precipitation levels, increasing temperature, and the loss of summer fogs. Summer fogs, in particular, are characteristic of the plant's habitat. Summer fogs provide additional summertime moisture and may be important to the germination and survival of its seedlings. Again, we are confronted with the difficulty of undoing Anthropocene effects on other organisms. Even when we agree on the desirability of taking action, it may be too late, and we often don't know enough about these plants and animals to act appropriately. Today, though, there is hope for this particular species as the plants developed from cuttings from the Franciscan manzanita, rescued "ghoulishly" by Lester Rowntree, are planted near the last wild Franciscan manzanita so that they may cross-pollinate, bringing this marvelous plant back to the hills surrounding San Francisco Bay.

The American burying beetle, an insect about 1.5 inches long, which doesn't bite or sting or damage property, is a threat to the

United States military—at least according to members of the United States Congress. Congressman Frank Lucas of Oklahoma in 2016 and Oklahoma congressman James Bridenstine in 2017 each introduced amendments to the National Defense Authorization Act, a Department of Defense budget bill, that would permanently remove the American burying beetle from endangered species status. The amendments claim the American burying beetle is "an unnecessary burden on national defense installations." The Endangered Species Act provides exemptions for national security, however, and the Department of Defense didn't request the amendment. The Department of Defense has actually assisted in conserving the beetle, found on a National Guard training facility, and at an ammunition plant in Oklahoma—the two locations that may have prompted the amendment. So who really benefits from the amendments? Why did the American burying beetle merit special mention in the defense budget bill? In explaining the amendment, the *Tulsa World* newspaper says that the beetle is a source of irritation to oil and gas operators, real estate developers, and road builders. It is not that the insect is physically irritating—it's more emotionally and financially irritating. Because the beetle is an endangered species, these developers need to mitigate the potential impact of their activities on the American burying beetle, which can interfere with their plans. It must be an onerous burden, because Senator James Inhofe, also of Oklahoma, introduced a specific bill aimed at the beetle, the American Burying Beetle Relief Act of 2014 (S. 2678). A "lack of mitigation options for developers" was an important consideration in the bill. Fortunately for the beetle, the bill was not enacted. How has a beetle that spends much of its time underground achieved celebrity status in the U.S. Congress, with a bill solely devoted to it?

In the 1980s, entomologists noticed that it had been a long time since anyone saw or captured an American burying beetle. A survey of museums determined that American burying beetles had disappeared in much of the eastern United States by the 1920s, and that the last specimens in most museums were found in the 1940s. Aside from only a few specimens found sporadically in other locations, the American burying beetle was known to occur only on an island off the coast of Rhode Island, and in a single county in Oklahoma. The beetle was once widely dispersed in eastern North America—found in thirty-two eastern states and three Canadian provinces. After learning of the beetle's precipitous decline, the U.S. Fish and Wildlife Service acted in 1989 to list the American burying beetle as an endangered species. Subsequently, with increased focus on the insect, additional small, localized populations were found in Nebraska, South Dakota, Oklahoma, Kansas, Texas, and Arkansas. Even today, no one is certain what happened to the American burying beetle. Speculation on the beetle's decline has focused on disease, pesticides, habitat fragmentation, an increase in competitors, and a decrease in the carrion they feed on. But no one really knows for certain what caused American burying beetles to disappear from most parts of their extensive range.

The American burying beetle is the largest representative of the carrion beetles in North America. It has a shiny black body with bright orange patches on its back, the front of its face, and at the tips of its antennae—a rather handsome representative of the carrion beetle clan. Other carrion beetles in North America have maintained their distributions and do not seem to be in decline. Whatever is affecting the American burying beetle numbers is peculiar to that particular beetle. Because of its larger size, the American burying beetle requires larger carcasses than the other carrion beetles, generally preferring small to medium-sized birds and mammals. Some scientists have even connected the decline of the American burying beetle to the extinction of the passenger pigeon. Their ranges would have overlapped, and the passenger pigeon would have been an appropriate-sized carcass for the American burying beetle. The vast numbers of passenger pigeons could have been sufficient to support burying beetle populations. More likely, though, there were additional factors involved in the beetle's decline.

American burying beetle adults can detect a dead animal soon after it dies, and from as far away as two miles. Early detection and speed are of the essence for the burying beetle, which must arrive at the carcass ahead of other scavengers. American burying beetles are nocturnal. Arriving at animal remains at night gives the beetles time to bury the dead animal before carrion flies begin to arrive in the morning. Upon arrival at the corpse, beetles may fight over possession. A victorious male and female cooperatively bury the carcass. If the ground isn't suitable for digging, they may have to move the body to a better location. The beetles achieve the relocation by crawling under the corpse and then propelling it with their feet while lying on their backs. They may move a carcass as much as several feet using this technique. Once satisfied with a location, the pair of beetles will remove soil from beneath the carcass, effectively lowering it below the ground surface, and then covering it with the soil that is pushed up the sides of the hole in the burial process. The male and female next remove the hair or feathers from the corpse, roll it into a ball, and smear antibacterial secretions over it. The female then constructs an underground chamber above the carcass, in which she lays her eggs.

The American burying beetle is unusual in the insect world in that both males and females participate in caring for their young. The adult insects feed on the carcass, and then provide regurgitated food to the larvae. The caterpillar-like larvae spend about a week with their parents, feeding on the carcass. When the larvae reach the requisite large size, they leave their parents and what's left of the carcass, and crawl away to pupate in the soil. The pupal stage lasts about a month, after which the newly formed adult beetle emerges from the soil, resplendent in its black and orange coloration. These adult beetles will subsequently spend the winter underground. The parents leave the nest around the same time as the larvae and die shortly afterward. American burying beetles live only a year, ultimately devoted to the task of raising new members of the species.

Although no one can pinpoint the factors leading to the loss of American burying beetles through much of eastern North America, it is likely that Anthropocene changes precipitated their demise. Some scientists have suggested that the fragmentation of North American habitats, and the elimination of large predators such as wolves and bears, may favor smaller, more adaptable mammals like raccoons, opossums, and coyotes, resulting in more competition for carrion for the American burying beetle. The destruction of woodlands, and other native habitats, may in itself have had a negative effect on the beetle. Dr. Wyatt Hoback, an entomologist and expert on the beetle at Oklahoma State University, says: "Like an elephant or a lion it needs a lot of space. Each beetle requires one hundred acres to survive." This kind of unpaved, unadulterated space is in short supply in the Anthropocene. Also, increased nighttime artificial light may disrupt beetle navigation. Ending up in the wrong place can be fatal for insects like these beetles, through greater exposure to predators or landing in an inhospitable environment like a building or someone's home. As we have seen with the other examples of endangered species in this chapter, the biology of these organisms can be quite complicated, and we may misunderstand the factors that contribute to the decline of populations.

Nonetheless, in each case, efforts have been made to undo the damage we have done to these species. That in itself is a remarkable change in our way of thinking about nature. We seem to have developed a greater sensitivity to the plight of other living things, extending even to the smaller things like dragonflies and beetles. Despite the protestations of Oklahoma congressmen, there is something noble about adapting human activities to the needs of an endangered beetle that has little apparent economic value. Hoback says that the "problems the beetle faces will accelerate without protection."[3] It is no longer found in Texas, and only small populations remain in four counties in Kansas and three counties in South Dakota. Reintroductions have failed or had limited success in Ohio, Missouri, and Massachusetts. These efforts continue and will hopefully yield self-sustaining populations in the future.

Unfortunately, our more noble instincts may not be prevailing in the case of the American burying beetle. As of May 1, 2019, the U.S. Fish and Wildlife Service has proposed downlisting the American burying beetle from endangered status to threatened. Additionally, the bureau plans to implement a 4(d) rule, which allows many land uses that might affect the beetle to proceed without regulation. In the process, Dr. Hoback, who was asked to serve as a scientific expert on the review, claims that published scientific data was ignored. The Fish and Wildlife publication on the proposed rule and request for public comment makes the odd assertion that although this endangered species is no longer endangered, it will be endangered again soon.[4] We are left to wonder why we are not taking action to prevent that rather than allowing developers and the petroleum industry to diminish American burying beetle habitat. It seems that moneyed interests may win a political victory over the American burying beetle after all. But if one political administration can change Fish and Wildlife rulings, then perhaps we can support another administration that's less cozy with these industries and more friendly to the beetles.

Several hundred years ago, most people, including scientists, did not think extinctions were possible. Our ability to conceive of extinction was limited by a static understanding of creation— the plants and animals present on earth had been there since the beginning. It seemed even less likely that human beings could directly cause the extinction of another species. Yet it was a little more than three hundred years ago that sailors hunted the dodo to extinction on the island of Mauritius. And, finally, it has been a little over two hundred years since the concept of extinction was accepted by science. Now, many scientists believe we are entering an Anthropocene extinction event generated by human activity.

Our understanding of biology changed significantly with the development of the theory of evolution. We now recognize that species come and go with regularity in nature—a possibility faced even by our own species. Nature is "red in tooth and claw," according to Tennyson, and we accept that species must constantly adapt to changes in their environment to survive.[5] Entomologists speak of an evolutionary arms race, in which plants develop new chemical defenses against the insects that feed on them, and insects evolve metabolic countermeasures that allow them to detoxify those chemicals. But human-caused extinctions somehow seem different. The Endangered Species Act in the United States refers to the "esthetic, ecological, educational, historical, recreational, and scientific value to the Nation and its people" of endangered species. The Endangered Species Act suggests that we should take action when we can prevent extinctions resulting from "economic growth and development untempered by adequate concern and conservation."[6]

Decisions to conserve species, and their habitats, become more difficult the more that a species is inconvenient to us. Are we willing to slow down in certain stretches of roadway to save the Hine's emerald dragonfly? Or does that idea seem rather comical? Are we willing to support more expensive oil and gas or real estate development to preserve populations of the American burying beetle? Moving further along the scale of inconvenience, should we try to eliminate the mosquito species that transmit malaria? We can agree that there might be a difference in organisms that are a danger to humans, such as the bubonic plague bacteria, as opposed to organisms that are an economic inconvenience. But there is also an ethical aspect to these questions. We will discuss this further in the final chapter, but here consider that the lives of other organisms may have value just as we believe human lives have value—and not just in their instrumental value to humans. The Endangered Species Act refers to the value of endangered organisms to the nation and its people. But does an organism that has survived for several hundred million years not have a value of its own that is independent of humans? Don't we have a responsibility to mitigate the impact of our actions on other species irrespective of whether they are useful to us?

Beyond remediation, and survival of a species, can we conceive of responsibility for the flourishing of a species? We don't know enough about the biology of many organisms to determine how best to help them. The large blue butterfly and the American burying beetle are examples of how difficult it can be to identify the needs of a specific organism. But we can still reclaim and preserve natural habitat, which must be a first approximation of providing what allows a species to flourish. The work done by land trusts and conservation societies plays an important role in these efforts. We must also recognize that the loss of one species may have a cascading effect on others. It is impossible to know the degree to which the extinction of the passenger pigeon or the loss of larger predators across much of the eastern United States affected populations of the American burying beetle, but it seems likely that these changes had an impact. The flourishing of a species is in part dependent on the presence of other species that, perhaps indirectly, create the requisite conditions for flourishing. We must pay attention, to monitor the health of our ecosystems so that we can maintain the conditions necessary to support other organisms. Perhaps we can all slow down a bit, and allow the Hine's emerald dragonfly to dart out of harm's way.

SIX

Human Exceptionalism?

THE IDEA THAT HUMANS are somehow superior or more valuable than other organisms is referred to as human exceptionalism. The Judeo-Christian tradition says that humans are made in the image of God. For Hindus and Buddhists, humans are at the apex of reincarnation for beings on this planet. Philosophers recognize human rationality, culture, and language as evidence of human superiority. Human beings in the Anthropocene are reshaping the planet, modifying the climate and geochemical cycles, and changing the conditions for life as we know it. The accumulation of carbon dioxide in the atmosphere is changing the earth's climate and contributing to acidification of the oceans. Industrial agriculture removes nutrients from the soil faster than they can be replenished. And when excess nutrients are applied, much runs off into waterways, changing their chemistry and resulting in areas like the dead zone in the Gulf of Mexico. More than any other time in our existence, the changes we have made to the planet over the past century, for better or worse, give us reason to believe in our exceptionalism. But can our judgment of what is exceptional be impartial? We naturally have an elevated opinion of human activities and behavior. But what if we took a more objective view? What if we looked at humanity and nature not from a human perspective but from the view of a visitor to our planet? That visitor would see that humans are remarkable creatures, who are indeed changing the world—though not always in a way that is beneficial for other living things, or even ourselves.

Termites, however, are also changing the world. An extra-terrestrial visitor would observe termites (genus *Odontotermes*) in Africa that are amazing engineers and architects; they build large mounds, some as large as eight meters tall. So termites are constructing structures that are about one thousand times larger than the termite itself. The tallest human structure as of this writing, the Burj Khalifa in Dubai, at 830 meters high, is merely five hundred times larger than a human being. These African termites are also great horticulturalists, culturing a specific fungus in their mounds for use as food. The fungus is fussy. To grow well, it requires high humidity, a constant temperature of about thirty degrees Celsius, and low levels of carbon dioxide. Termites construct their mounds to meet those needs. The temperature inside large fungus-growing termite mounds in Africa varies by only plus or minus two degrees Celsius, even though temperatures outside the mound vary by as much as thirty-five degrees. That's pretty remarkable climate control. Furthermore, fungus-growing termites modify the structure of their mounds depending on location—either to retain heat in cooler forest environments or to provide more cooling in the hotter savanna. The termites control airflow and carbon dioxide exchange in their nests by developing air channels in the mound, controlling the thickness of mound walls, and modifying the surface area of the mound itself. Farming this fungus and rearing young in a large, climate-controlled structure that houses millions of termites is all accomplished without using fossil fuels. So termites are exceptional architects and builders—in many ways perhaps as exceptional as humans.

Termite structures are important not only for the termites themselves, but also for their ecosystem. In the savanna, termite mounds and the soil around them are more porous and thus retain rainwater better than the soil beyond them, which is important for local plants. Termites may add clay to soil that is too sandy for building sturdy tunnels or may add sand to hard soils difficult for tunneling. In doing so, they alter the soil in ways that are advantageous for plant growth. Termite mounds therefore encourage the development of communities of plants, plant-associated insects, other plant-eating animals, and the animals that eat the insects.[1] The favorable plant habitat provided by termite mounds also serves to prevent or slow desertification in arid areas. The mounds termites develop act as a sort of plant oasis. During dry periods, plants may be limited to growing on or around these oases. When the rains come, plants can expand their distribution away from the mounds. But when termite mounds are removed and the land is cultivated by humans, it becomes much more susceptible to desertification. Termite construction is integrally involved in and even contributes to the ecosystem. They set a standard to which we can aspire in our green LEED and Living Building initiatives.

The argument for human exceptionalism is grounded in our rational, societal, and tool-using capacities. We can argue that the success of the human species is evidence that we are exceptional.

But what do we mean by success? How would you define success for other organisms, such as termites, ants, bacteria, or dung beetles? Humans have been successful in manipulating their environment to ensure the perpetuation of their species. But termites also successfully modify their environment, and in ways that are ecofriendly—perhaps that is a more universal success since it benefits other organisms. Success might mean something different for humans than for ants. Is success related to numbers? If so, the ants win. Is it related to geographic conquest? Ants probably win that one, too. Could longevity of the species be evidence of success? Most insect species predate *Homo sapiens.* And what of physical and mental capacities? It turns out that our physical and mental functions are not, strictly speaking, solely our own. The bacterial cells that reside within our bodies play a significant role in our metabolism, and affect our thought and emotions. What we think of as human results from a combination of the efforts of bacterial and human cells in our bodies. Is there an objective reason to value human capacities over those of insects? Gram for gram, dung beetles are far stronger than humans, and their activities are beneficial for the environment and us. Human Anthropocene activities would seem to be deleterious for the environment, rather than contributing to it as dung beetles do. Could it be that the abilities we value so highly are not important to other forms of life? Considering human abilities from an ant's standpoint might be a little humbling.

By many measures, the ants are a spectacularly successful group. The earliest fossil records of ants suggest that they first appeared over 100 million years ago. The earliest *Homo* species, human ancestors, appeared a little less than 3 million years ago, and *Homo sapiens* has walked the earth only over the past several hundred thousand years. Humans have certainly not achieved the long-running evolutionary success of ants, and we would need to maintain our existence for hundreds of thousands of years in the future to do so. Ants also outnumber human beings. Estimates of human population put us just over 7.5 billion individuals in 2019. Estimates of ant numbers vary widely, but they may outnumber humans by more than a million to one. Instead, you may argue that humans are a lot larger so their numbers count for more than a single ant. Of course, this is true, but if we were to weigh all ants and all human beings, we would find that the biomass of ants (or their total weight) is ten times that of humans. Although these numbers may be somewhat surprising, our sense of superiority is probably more related to human intelligence.

At first glance, human brainpower far exceeds that of insects. Human beings are estimated to have at least a hundred thousand times more neurons than insects do (using honey bees as a standard). But scientists have discovered that ants have cognitive capabilities beyond what one might expect: they are not mere automatons. Ants are able to learn. They can remember locations of food, the time food in that location is available, and how to find their way home. Ants may use memorized landmarks in addition to calculating distance and direction on their trip. Some ants have also been observed using tools. Several ant species use bits of soil, leaf, sand, or pine needle to adsorb and carry liquid or semi-liquid food back to the nest. Researchers have observed that ants could carry more material on the tool than they could carry internally (by ingesting it—another way ants carry food back to the nest). Other ants use tools as weapons. Several ant species will drop soil particles or pebbles into other ant nests to prevent competitors from foraging. Some ants drop pebbles into ground-nesting bee holes, and then attack and kill the bee when it emerges.

The renowned biologist Edward O. Wilson suggests that we should think of an ant colony as a superorganism rather than as a collection of individuals.[2] He reminds us that natural selection acts on the colony itself, rather than on individual worker or nonreproductive ants. Characteristic organization of the ant superorganism includes female nonreproductive worker ants that perform a variety of tasks in the colony, including foraging, care of young, care of the queen(s), and defending the colony—these workers may be further divided into castes that perform particular tasks. Ants cooperatively engage in these tasks, which are necessary for supporting the colony. The queens and males are the reproductive members, although the males are generally present only during the breeding season. Communication among the ants in a colony is conducted via chemicals. The superorganism simultaneously gains information about its environment through the many worker ants that conduct multiple forays outside the nest identifying the location of food, or on occasion potential new nest sites.

A study of the ant *Leptothorax albipennis* revealed a sophisticated selection process involved in choosing a new nest site, when the original site was damaged. Choice of a new nest site required site quality assessment by multiple ants. The decision to accept a new nest site is made jointly by colony members. The authors of the study refer to this as a sort of "opinion polling."[3] In this example, you could say that the colony, as a superorganism, integrates sensory information, and then makes a decision based on this information. Not unlike how a human being chooses a new home. The ants must consider the relative quality of sites as well as how much time to devote to the search process. Searching too long when the original nest site is compromised may be disastrous for the colony. The cognitive capabilities of ants should be judged more on the cognitive capability of the colony rather than on its individuals. Individual ants may have a limited set of behavioral capabilities, but the colony as a whole has a much broader behavioral repertoire, combining the actions of all workers and reproductive individuals to feed, maintain, protect, and reproduce the colony.

The Argentine ants (*Linepithema humile*), found in North America, Japan, and Europe, appear to be so closely related as to form a giant supercolony of ants. These ants originated from the floodplains of northern Argentina and southern Brazil. In their native range, Argentine ants live in smaller colonies, constantly competing with other Argentine ants and ants from other species. Ants from separate Argentine ant colonies are aggressive toward one another. As in the case of the large blue butterfly, ants from the same nest recognize each other by odor. Argentine ants attack ants that have an odor that differs from that of their colony. These ants from South America were introduced in New Orleans in the late 1800s and reached California by 1907. Scientists believe that the introduction into California from a single source resulted in a limited genetic profile for Argentine ants there. As a result, workers from new nests produced by that colony continued to recognize each other as nest mates, having a similar odor. They are not aggressive toward each other. The Argentine ants in what is essentially a single colony in California, based on nonaggression between workers, now extend for over five hundred miles along the coast. Subsequent introductions of Argentine ants have resulted in additional supercolonies in California that compete at the edges of this largest supercolony.

In Europe, a similar Argentine ant supercolony extends along the Mediterranean Sea for more than two thousand miles. Researchers have found that Argentine ants from supercolonies in Japan, New Zealand, Australia, Hawaii, Europe, and California behave toward each other with little or no aggression—as though they were from the same colony. Nonaggression between nests within the supercolony may contribute to its success, because the ants don't need to expend energy protecting a territory. It is amazing to conceive of the supercolony as a single superorganism spread around the world. The globally distributed Argentine ant superorganism is simultaneously processing and responding to environmental information on multiple islands and at least four continents. This is distributed knowledge and behavior— there is no central processing faculty in the superorganism. Knowledge of the environment or status of the colony is developed through the interaction of individual ants, each of which obtains a portion of the environmental information. Human beings are also social creatures, but they cooperate in looser, more malleable ways than ants. And as individuals, humans have far greater cognitive capacity than individual ants. But ant societies are just as successful biologically as human beings, using a different cognitive approach—a more collective intelligence.

Let's consider next another sort of "bug" that is probably even more biologically successful: bacteria.

Bacteria are found in a far broader range of environments than either humans or ants. The community of microorganisms in a particular environment is often referred to as a microbiome. Scientists have studied the microbiomes found in the atmosphere, miles above the earth's surface, to the deepest part of the oceans—under the cold, dark, and high-pressure conditions found more than thirty-five thousand feet deep in the Mariana Trench of the Pacific Ocean. Bacteria are also present in lakes formed more than two thousand feet below the surface of the ice in Antarctica and in Yellowstone National Park hot springs, where water temperatures approach boiling point. Aside from the bacteria found in these extreme situations, bacteria are universally present in soil, water, air, on the surface of living things, and inside of them.

We humans tend to think of ourselves as something apart from nature. In fact, nature is often understood as that which is not human created or human influenced. Our genetics and evolutionary history belie this idea. We're related to a greater or lesser degree to the full range of life, and share common ancestry going back to the earliest single-celled forms of life. A favored explanation for the evolutionary development of our cells, eukaryotic cells, is symbiotic bacteria taking up residence inside another single-celled organism—most likely archaea, or a common ancestor to archaea and eukaryotes. The evolutionary result was the transformation of those symbiotic bacteria into mitochondria, the powerhouse of the cell, which provides the energy needed for the evolutionary development of the expansive range of living forms made of eukaryotic cells—including fungi, plants, and animals. So each human cell contains an organelle whose origin traces back to what might have been a parasitic (it may not have been symbiotic in the beginning) bacterium. Not only that, but human beings are also an ecosystem in which smaller organisms live their entire lives on or inside of us. In fact, microbiologists have discovered that humans have a microbiome of their own—and it has become a hot research topic. In the United States, the National Institutes of Health includes a Human Microbiome Project, and there is an International Human Microbiome Consortium that coordinates the human microbiome research efforts in multiple countries.

The website for the Human Microbiome Project reports that there are ten times more bacterial cells than human cells in the human body. Many, if not most, of these bacterial cells are integrally associated with how our bodies function. We are only beginning to understand the different ways that our microbiome is associated with our health. It is responsible for production of certain vitamins that we can't make ourselves, producing enzymes that aid in digestion, assisting our immune system, and protecting us from bacterial and other toxins. Scientists are now learning that chemicals produced by bacteria in the human gut may affect our brains. Bacteria in the gut produce many of the same chemicals that neurons in our brain produce. Experiments in mice appear to confirm that chemicals produced by bacteria in the gut affect behavior. Mice that are raised bacteria-free, for example, suffer deficits in memory and sociability, and demonstrate increased locomotion, self-grooming, and anxiety. In humans, the brain communicates with the gut and vice versa. Communications between the brain and gut bacteria can take place through the nervous system, through hormones, and through immunological interactions. Gut bacteria appear to play an important role in brain development in children. In addition, researchers have discovered an association between the composition of gut microbiome and temperament in children. Scientists also speculate that the makeup of the gut microbiome may influence conditions such as anxiety, depression, and autism. We will certainly learn more in the future, as this field of study is advancing rapidly, but it's clear that human mental function is linked to our associated bacteria.[4]

It's a little disorienting to think that who I am is related both to my human genome and to the multiple bacterial genomes in my body. In fact, my bacterial genes outnumber my human genes by about a hundred to one. Those bacterial genes actively program chemicals involved in internal messaging and metabolism. In that sense, I may be more bacterial than human. How does this affect our image of human exceptionalism? If we are exceptional, it's the product of teamwork between our human and microbial symbiont genomes. Thus we are far more intimately related to non-human nature than we might suppose. It seems that our microbiome influences our thoughts, feelings, and behavior. In a real sense, therefore, our microbiome is part of who we are. This connectedness to non-human nature contradicts notions of humanity as a separate entity somehow outside nature. Instead, we are embroiled in nature, both in our homes and in our bodies. We are less exceptional and more a unique part of a natural continuum of life. Our microbiome betrays a deeper relationship to the rest of life. The microbiome itself varies among individuals: the composition of one person's microbiome may differ considerably from another's. The gut microbiome's composition changes as a person ages as well, with dietary changes, stress, infections, and use of antibiotics. The gut microbiome may have 100 trillion bacteria, consisting of a thousand different species. But there is a set of 150–200 core bacteria species commonly found in the human gut. Who are these frequent inhabitants of humans?

Bacteria of the genus *Bifidobacterium* are dominant in colonizing the human infant gut. *Bifidobacterium* species may be passed to infants in mother's milk. It has been shown that some *Bifidobacterium* and *Lactobacillus* species produce a neurotransmitter chemical called GABA (gamma-aminobutyric acid). *Lactobacillus* species also produce acetylcholine, another neurotransmitter chemical in the body. *Lactobacillus* is generally the first bacteria to which infants born vaginally are exposed. Both of these bacteria are speculated to be important to early brain development in infants. *Bifidobacterium longum* is a rod-shaped bacterium that lives under anaerobic conditions (that is, without oxygen). As we age, *Bifidobacterium longum* drops from 90 percent of the gut microbiome to 3 percent, and is replaced by a diversity of other bacteria, such as *Bacterioides* species, which become dominant in the adult gut. *Bifidobacterium longum* has been connected with a variety of health benefits as a probiotic—a bacterium that is beneficial to the body. Beyond physical health, there has even been discussion of psychobiotics—bacteria that have a beneficial effect on mental or emotional health. Clinical studies will be required to verify mental health effects, but as we have seen there are studies that support the idea that bacteria influence our mental and emotional well-being.

Bacterioides thetaiotaomicron, another rod-shaped bacterium that lives only in nonoxygenated (anaerobic) conditions, is well adapted to life in the human colon and flexible in metabolizing the variety of foods we may consume, from nachos to gazpacho. It serves as an example of how hard-working bacteria benefit human beings. This bacterium plays an important role in nutrient availability for other microorganisms in the gut, and for human beings. *Bacterioides thetaiotaomicron* is able to break down complex molecules that would be otherwise indigestible for its human host. This is important because 10 to 15 percent of human calories are derived from microbial fermentation in the gut. And this bacterium works as a partner there, communicating with host intestinal cells, and even influencing gene expression and the production of chemicals by those cells. Finally, *B. thetaiotaomicron* contributes to the immune response in the human gut. The bacterium may encourage host intestinal cells to produce antibacterial compounds that prevent infection by pathogenic bacteria. We humans (and other organisms too) depend on relationships with bacteria that have evolved over long periods of time. Our microbiome influences who we are and how our bodies function. Rather than saying you are what you eat, given what we now know about our microbiome, we should say, you are what they eat. Here we find further support for the growing recognition of the powerful influence of small living organisms.

The lowly dung beetle is another smaller organism that is more powerful than you might imagine.

The biologist J. B. S. Haldane is reputed to have said that God must have had an inordinate fondness for beetles. The exact wording may be apocryphal, but Haldane's point about beetles is accurate. We have identified around four hundred thousand species of beetles, and there may be at least ten times that many that haven't been identified. Beetles are thus the most species-rich order of animals. Scientists have suggested that part of the reason for the enormous number of species is a low rate of extinction. Beetle species survived several major extinction events, including the one that killed all the dinosaurs. Dung beetles in the scarab family are a particularly interesting example, as it is believed that they evolved eating dinosaur dung. These scarab dung beetles may have evolved as many as 150 million years ago, when dinosaurs began incorporating the newly available flowering plants in their diet and dinosaur dung became more palatable. It appears that some of the most narrowly focused early dinosaur dung-feeding specialists may have been lost with dinosaur extinction. But other dung-feeding beetle species with more flexible food habits survived, and new species developed as new plant-feeding mammal species arose.

Dung beetles received quite a bit of attention a couple of years ago when it was ascertained that the bull-horned or bull-headed dung beetle (*Onthophagus taurus*) was the strongest animal, relative to its weight, in the world. The greatest ratio of weight moved to body weight by a human being was established when 364-pound Paul Anderson, known as the world's strongest man, is said to have raised 6,720 pounds on his back off of a trestle in 1957—a little less than twenty times his body weight. In contrast, the bull-horned dung beetle pulled 1,141 times its body weight in laboratory studies conducted in 2010.[5] Why has the bull-headed dung beetle developed such herculean strength? To understand that, we need to know a little more about the life of bull-headed dung beetles.

Bull-headed dung beetles get their name from long curved horns extending back from just behind the head in males. Males use these horns in combat with other males for access to a female. Generally, the male with the longer horns wins the battle—unless the horns are relatively similar in length. Then the slightly shorter horns may have an advantage. Oddly, some males are not so bull-headed, and are hornless. These hornless individuals are referred to as minor males, in contrast to the long-horned major males. The major and minor males engage in different reproductive strategies. The major males attempt to win and guard a female through aggressive tactics. But the minor males attempt sneak mating. The sneaky minor males have enlarged testes and are speedier walkers than the majors. When given the opportunity, the minor males can quickly inseminate a female, injecting a higher quantity of sperm in their sperm package than the major male. As a result, if the female mates a second time with a major male, the minor male's sperm stands a proportionately better chance of fertilizing the female's eggs. The minor males are opportunistic. They may slip past majors involved in battle, may catch a female when she emerges from her tunnel to collect dung, or may enter the female's tunnel from a side tunnel. Though less noble, sneak mating can be a successful strategy for minor males.

As you may have gathered, much dung beetle activity takes place underground. Bull-headed dung beetles construct a tunnel extending several inches below a dung pat. The female dung beetle creates small brood chambers extending off the tunnel, which she provisions with an elongate ball of dung. Major males may assist in the dung provisioning process if not too busy guarding or fighting off rivals. The female lays a single egg in each ball of dung, and the young or larvae complete their full development there. Then, the newly developed adult beetle emerges from the tunnel, locates another dung pat using olfactory senses, and begins the cycle again.

Dung beetles provide several valuable ecological services. If it were not for dung beetles, and other dung-feeding insects, we would likely be waist deep in manure. In fact, human beings have imported dung beetles into numerous locations around the world to control the accumulation of cattle manure. The importation of dung beetles into Australia is a good example of this. Native Australian dung beetles evolved with marsupials, which produce hard, dry dung. The beef cattle imported into Australia tend to have softer, more liquid dung that does not attract native dung beetles. The growing surplus of cow dung pats that resulted in Australia was causing two immediate problems. First, understandably, cattle refused to feed in fields filled with cow dung. And the cow dung resource was also contributing to a banner crop of bush flies (*Musca vetustissima*), which are very annoying to people, often landing on their faces in search of moisture around the eyes, nose, and mouth. The prevalence of bush flies led to what is now called the Aussie salute—a waving motion employed by Australians to shoo bush flies away from their faces. You can imagine the cow dung situation was rather dire.

The problem was resolved by importation of a number of different dung beetle species that efficiently process cow manure. Not only do dung beetles remove dung pats, but they also play a significant ecological role: the tunnels produced by the bull-headed dung beetle aerate the soil, and the dung brought into the tunnels contributes to soil fertilization. Furthermore, dung beetles may also reduce methane that contributes to climate change. Dung from livestock, particularly cattle, is a significant contributor of greenhouse gases that cause climate change, especially methane. The aeration and burial of dung pats by dung beetles causes a significant reduction in methane emission in areas where cattle graze. Dung beetles are not only the strongest of animals on a pound-for-pound basis, they are essential to the functioning of natural and agricultural ecosystems.

I asked at the beginning of this chapter whether the human species, *Homo sapiens*, is somehow superior to other species. Are humans exceptional? In considering the lives of other successful organisms, I think we have discovered that, although human beings are unique in several ways, we should not understand them as superior. Determining exceptionalism or superiority entails a judgment about what is important. Perhaps we humans have a biased view of the value or importance of what separates us as a species. It is tempting to look at how we control and modify our environment as evidence of superiority. The view of the earth from an airplane testifies to the impact of our species, as we pass over acres of farmland, towns, cities, and suburbs. Yet, as we have seen, other organisms modify their habitat, construct their own homes, and live just as expansively on the earth.

We are accustomed to seeing life depicted as a graceful evolutionary tree, with human beings envisioned at the tip of a branch at the top of the tree. The protozoans, bacteria, and archaea form the base or part of the tree trunk in this depiction of evolution. Biologically, however, this is not a realistic portrayal of the evolution of life. Instead, the most recent evolutionary diagrams look like a big, round shrub, mostly composed of different bacterial lineages. A small branch in the shrub of single-celled life represents the multicellular plants and animals that are visible to us—what many of us think of as nature. Although we often speak of the age of dinosaurs or the age of mammals, an objective view of life on earth would recognize that we have always lived in an age of bacteria, which occupy the full range of environments on the planet in numbers far beyond any other kind of organism.

Just as humans are altering the earth's atmosphere today in the Anthropocene, bacteria more than 2 billion years ago contributed oxygen to the atmosphere, changing it to something more like what we experience today. Just as the Anthropocene appears to be the basis of a great extinction event, the oxygenation of our atmosphere by bacteria was also a great extinction event, eliminating countless species of anaerobic bacteria—organisms that don't use oxygen for growth and for whom oxygen may be toxic. Bacteria are pervasive on the planet, and live in and on plants and animals, often in symbiotic relationships. Many species of plants and animals could not survive without their bacterial symbionts. Humans too, if we could survive, would be different creatures without our microbiome. The tree of life, as we understand it today, places human beings in perspective—a life form, like other multicellular organisms, dwarfed by the profusion of bacteria. Rob Dunn, in his book *Never Home Alone*, speaks to the omnipresence of bacteria in human environments. In his view, we benefit from the diversity of bacteria found in our homes and should embrace it rather than trying to eliminate it.[6] But we also have mentioned other ways of thinking about human superiority that exclude bacteria—functions that only a multicelled organism can accomplish.

We laud members of our species for feats of speed and strength. Yet we are relative weaklings compared to other species. If we admire strength, then we should award an Olympic gold medal to the dung beetle as the strongest animal and perhaps for the critical ecosystem services it provides. We can't really argue that our social organization makes us exceptional, since ants have a noteworthy and effective social structure. The ant social structure is different from ours, but it has contributed mightily to their biological success. Here perhaps we should pause a moment and consider what success might involve for a species in biological terms. The perpetuation or success of a species over long, evolutionary-scale periods of time is the result of continued reproductive success, which is conditioned by an organism's adaptability. The organisms we have considered in this chapter have demonstrated long-running evolutionary success. Dung beetles have been around for more than 100 million years. Ants evolved about 100 million years ago. A group of bacteria called the cyanobacteria evolved around 3 billion years ago. In contrast, human ancestors made an evolutionary split with chimpanzees about 8 million years ago, and *Homo sapiens* appeared less than 500,000 years ago. We have a long way to go before we can claim the evolutionary staying power of dung beetles, ants, or cyanobacteria. There is no doubt that human beings are very successful organisms, but can we maintain that success over evolutionary time? Perhaps our cognitive capabilities will contribute to longer-term success, but it seems they might also contribute to our downfall.

Humans are using those cognitive capabilities in an ongoing battle against insects, plants, and other organisms. Entomologist May Berenbaum suggests that insects become pests because they like the same things we like—particularly our food and shelter.[7] We have created a variety of ways, mostly chemical, to eliminate the living things that compete with us for food, space, and shelter. Life, though, responds to human depredations. Insects, plants, fungi, and others evolve strains that are resistant to pesticides, resulting in an escalating arms race. Humans assault with ever more clever chemical structures, and still new resistance develops.

I observed a fascinating case study while working on pesticide products myself. I spent a number of years working on insect bait products. Many of us hoped that these would replace or at least reduce the use of pesticides sprayed in broadcast fashion against ants and cockroaches. Baits were formulated with insecticide and food ingredients, like sugars and proteins that would stimulate feeding by the insect, and thus ingestion of the killing ingredient. The insecticide-laced food bait is contained in a plastic receptacle, with small holes that allow insects to enter, but prevent access by pets and children. Furthermore, the most successful baits used new insecticides that didn't kill immediately, and allowed the insect to return to hiding places or the colony where additional insects would be affected by the chemical. These insecticides were chemicals that acted in new ways on insect metabolism, and it was thought that it would be much more difficult, if not impossible, for insects to develop resistance to them. Instead, we learned to never bet against the creativity of life. Entomologists were puzzled to find that, after spectacular success in exterminating cockroaches with these baits, the products began to fail. Could cockroaches have developed resistance to the new insecticides? Initial investigations found only low levels of resistance, which was even more confusing. In a change more fundamental than most believed possible, scientists discovered that the cockroaches changed their feeding behavior to avoid the sugars in the baits. In lab experiments, cockroaches would not eat a particular sugar found in the bait, even to the point of starvation.[8] Life and its evolutionary processes outwit human cleverness again.

Some scientists suggest that there is a basis for believing that even insects have a form of consciousness related to subjective experience—they are aware of events as happening to them as an individual. Scientists reason that the insect brain performs similar functions to the mammalian midbrain, the purported seat of subjective experience.[9] Our estimation of the cognitive power of other animals is made difficult by our inability to communicate with them. Even so, we are finding that many animals share some of our basic cognitive capabilities. Primates, dolphins, and elephants provide examples of animals that engage in certain behaviors regarding others, or something we might call moral behavior. We have also observed that even ants exhibit memory, tool use, and learning abilities. Outgrowths of our cognitive power, such as human culture (including writing, the arts, and the development of scientific understanding) and the extent of human tool making, are perhaps what set us apart. These are wonderful and unique qualities of our species that are unmatched in other animals. Do these capabilities mean that we are somehow superior to other organisms? Perhaps if we consider the question based not on what humans value, but on an evolutionary time scale, we would have to say that the jury is still out. And, as many have observed, our unique capabilities, which help us to explore the broader universe, have also created the tools of warfare that could lead to our destruction. It will take another hundred million years or so to know whether we can equal the evolutionary success of manure recyclers, like the dung beetles.

SEVEN

Coexistence

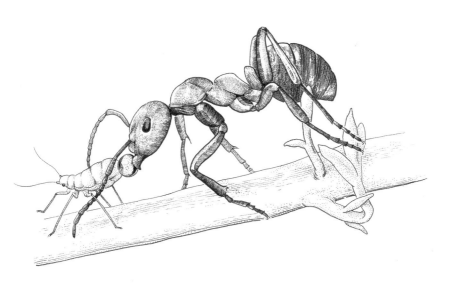

HOW FAR BACK can you trace your ancestry? A number of years ago, I took time to do a little genealogical investigation. I learned that my family, like many in the United States, is part of that great melting pot of nationalities. My maternal ancestors originated in England and Ireland. A paternal great-grandmother came from Hungary, then a part of the Austro-Hungarian Empire. My paternal grandmother was able to identify descendants back to the American Revolution. Just as we honor our grandparents, great-grandparents, and so forth, should we not honor our more distant relatives? Would it be odd to take that interest in ancestry, and our roots, beyond human beings to even earlier ancestors? It might be difficult to feel a connection with the shrew-like creature feeding on insects about 160 million years ago, which formed the first step in placental mammalian evolution. Or, going even further back to our greatest grandparents, the first single-celled organisms. We share portions of our genome and certain metabolic processes with all of life. Despite our similarities, we compete for resources as individuals within a species, and we may compete between species. An oft-repeated lesson from a popular understanding of evolution is that our world is a crucible of competition, expressed as "survival of the fittest." But the relationships among organisms are more nuanced and complex than that. There are certainly winners and losers over evolutionary time, but many organisms have developed cooperative relationships and feedback loops within their species or with other species.

There is much to be said about the value of common, smaller organisms and their importance to us. Appreciating the nature that surrounds us leads to an interest in conserving the diversity of life. But what approach do we take? The natural world is full of examples of feedback and regulated systems. Our bodies generally maintain a constant internal temperature of 98.6 degrees Fahrenheit. We sweat to cool ourselves and shiver to get warm. In addition to regulating an inner state of equilibrium, many organisms demonstrate responsiveness to their environment. The Old Testament tells us, "Go to the ant, you lazy bones; consider its ways and be wise" (Proverbs 6:6, NSRV). What can we learn from other organisms about cooperative behavior, and use of resources? Up to this point, we have been thinking about how human beings relate to other creatures. But perhaps there are lessons for human interactions with nature in how smaller plants and animals interact with each other.

Our human population has increased dramatically in the Anthropocene. It has grown from perhaps 5 million when we first began to practice agriculture to more than 7.5 billion people today, straining the resources of our planet. In contrast to our heedless reproduction, small, mouse-like animals called voles actually regulate their population size. Vole populations are known to fluctuate over time. Yet scientists could not identify what factors were causing this fluctuation. Vole populations have the potential to grow rapidly, producing multiple generations in a single year, but population growth was unpredictable.

To determine the factors affecting vole numbers, in this case the meadow vole (*Microtus pennsylvanicus*), scientists conducted an experiment, creating low, medium, and high populations of voles at separate sites. The scientists noticed that voles in the high-population sites started reproduction later and stopped reproduction sooner than the voles in lower-population sites. As a result, voles in the high-population sites had five fewer generations over the year than the voles in the low-population sites. Having fewer generations in the high-population sites would significantly reduce the growth of vole numbers. Thus voles naturally modify their reproductive behavior to avoid overpopulation. Other factors may contribute to regulating vole populations, of course, but their internal self-regulatory ability is important. Obviously, there are many more examples of animals that overrun their resources and whose populations are regulated by disease, predation, or starvation, but voles stand as a fascinating example of population self-regulation.

Humans have exceptional consciousness and reasoning capabilities, seemingly unmatched in nature. We too have recently made efforts to reduce our population on a crowded planet. In some instances, we delay the birth of our first child, and some families also limit the number of children they have. But because of our effect on the earth and consciousness of that effect, we have a greater responsibility than the voles—a responsibility to not only limit population growth, but also to limit our impact on the planet.

Self-regulation is one way that organisms coexist, but cases of self-regulation are relatively rare and have evolved to directly benefit only one species. Cooperative behaviors that may evolve between quite dissimilar organisms are another form of coexistence and are more common. Examples of these relationships include those of ants to plants, ants to aphids, and the insects associated with milkweed plants.

The aphids in your garden, like acacia plants (genus *Vachellia*), often have a beneficial relationship with ants. And though we may be proud of human efforts in bioengineering, we should note that aphids and their ancestors have been cloning themselves for hundreds of millions of years. The milkweed plant (genus *Asclepias*) plays host to a variety of insects. This is not a cooperative interaction, but one of evolutionary feedback loops. Many milkweed species produce toxic chemicals as feeding deterrents, yet their insect herbivores not only feed on milkweed, but also retain milkweed poisons to deter predators. These insects rely on the milkweed for both food and protection. Mutually beneficial relationships, clones, and chemical warfare are only some of the kinds of relationships living organisms form with each other. Ours is a dynamic world in which living organisms have used each other to further their own existence in manners both creative and bizarre.

One such cooperative relationship is that of the acacia ant (members of the genus *Pseudomyrmex* in Central America and the genus *Crematogaster* in Africa) and the acacia plant occurring in Central America and Africa. In Central America, queen acacia ants are attracted to the acacia plant by the plant's odor. On reaching the plant, the queen ant will hollow out a large acacia thorn, where she deposits her eggs, and a colony of acacia ants will develop. The acacia plant provides not only housing but also food for the ants. Small nectar-producing glands at the base of acacia leaves, called nectaries, provide highly nutritious liquid food for the ants. The acacias also produce food packets at the ends of their leaves that provide the ants with additional nutrients. The ants earn their room and board, though. They protect the plant from animals that try to eat acacia leaves by biting and stinging the intruders. They also remove other plants that grow near or over the acacia, which might reduce its access to sunlight and nutrients.

Scientists have demonstrated the importance of these ants to their acacias by experimentally keeping ants away from the acacias in exclusion studies. The antless acacias suffer defoliation by plant-eating animals and may be crowded by other plants that begin to grow over them and shade them. Ants thus contribute to the acacia plant's fitness. Similar experiments have also revealed that more microorganisms causing plant disease are present on acacia leaves when ants are excluded. The acacia ants appear to have bacteria on their bodies that reduce the presence of the plant diseases. The relationship does have a drawback for the plants, however, because they endure an associated cooperative arrangement. The acacia ant allows scale insects (insect family Coccidae) to attach to the acacia stem to drink plant sap. The scale insects in turn excrete sweet nectar, called honeydew, which ants also enjoy. So the ants, though an important partner for the acacias, are not completely reliable, carrying on a side job protecting scale insects.

Scientists in Africa have also used exclusion studies to better understand how ants and acacias benefit each other. In Africa, the exclusion studies allowed ants to reside on the acacias, but fencing prevented animals such as giraffes from grazing on the plants. The acacia plants within the fenced enclosure were then compared with plants that were not protected and thus accessible to giraffes. You would think that the protected plants would thrive with their acacia ants and no giraffe grazing. But it didn't work out that way.

What happened? Although acacia ants prevent the grazing of young giraffes, the older giraffes don't seem to be deterred by the stings and bites of the ants and continue to feed on acacias. So outside the enclosure, these older giraffes continued to graze on unprotected acacias. The scientists were surprised to discover that the acacias in the enclosed areas, though protected from giraffes, seemed to be less healthy than the acacias outside the enclosure that suffered giraffe feeding. When they investigated, they learned that the ant species that best protected the acacias, known as *Crematogaster mimosae*, had declined inside the enclosure. They were being replaced by another ant species, *Crematogaster sjostedti*. This second ant species, *C. sjostedti*, lives in holes in the acacia stem produced by a beetle from the insect family Cerambycidae. These beetles occur much less frequently in acacia trees tended by *C. mimosae*. Increased activity of the beetles, as *C. mimosae* dropped in numbers, and the holes they bored were likely the cause of the decline in acacia health.

Why did the numbers of *C. mimosae* wane and thus allow the wood-boring beetle and its associated ant, *C. sjostedti*, to become more numerous? Scientists determined that the acacia plants in the enclosure were producing less nectar and had fewer thorns that *C. mimosa* could use for nesting. *C. mimosa* success depends on the nectar and thorns, while *C. sjostedti* does not. It turns out that giraffe grazing appears to stimulate the production of both nectar and thorns in acacia plants. Without the giraffes, plants had less nectar, fewer thorns, and fewer *C. mimosae* ants—and thus more beetles and more *C. sjostedti* ants.

This story reminds us how difficult it is to predict the effects of changes we make in nature. It also demonstrates the extraordinary interconnection among community members—in this case a plant species, two ant species, a beetle, and the giraffe.

The relationship between ants and aphids is frequently used as an example of mutualism—both organisms benefit from the relationship. Aphids produce the sugary liquid honeydew that sustains ants, and ants protect the aphids from predators. But of course it's not quite as simple as that. There are many species of aphids that don't benefit from ant protection, and many species of ants that don't feed on honeydew. Aphids have long needle-like mouthparts, which they insert into leaves and stems to withdraw liquid plant sap from the tissue that carries it. The sap itself is kind of like junk food. It's high in sugar content for energy, but low in nitrogen for production of proteins. So aphids have to process a lot of plant sap to acquire enough nitrogen for growth. And, boy do they grow. Aphids are one of the fastest reproducing insects, producing as many as fifty generations in a year under optimal climatic conditions.

The Argentine ants described in Chapter 5 may tend one of these prolific aphids, the cotton aphid (*Aphis gossypii*). Not only do they have many generations in a season, but cotton aphid females in each generation may also produce as many as seventy to eighty young during their short reproductive life. In warm climates, such as the southern United States or in Hawaii, cotton aphids reproduce continuously throughout the year. Reproduction is generally not sexual in these climates. Like many other aphids, cotton aphids reproduce asexually, or parthenogenetically. Female aphids produce more female aphids. Males are not required. This strategy, along with short generation times and the production of many offspring by an individual, results in the extremely high reproductive rates found in aphids. Some aphid species' females contain a developing embryo, which in turn has another developing embryo inside it. Thus, the grandchild of the mother aphid is developing inside her daughter before the daughter is even born! It's an arrangement reminding one of Russian nested dolls. Males are produced, and sexual reproduction in the cotton aphid occurs in the fall of the year in more temperate climates.

The aphid reproductive strategy is certainly considerably different from human beings, with fifty generations in a year versus one every twenty-five years or so. Each individual aphid may produce as many as eighty young, whereas human females produce a little over two children on a global average. Predators, parasites, disease, and limitations of food resources keep aphid populations from overrunning the planet. Unlike aphids, humans possess a powerful ability to limit the depredations of disease and hunger, so we must exert some self-regulation over our relatively (compared to aphids) slowly increasing population.

Since aphids must process copious amounts of plant sap to acquire sufficient amino acids to manufacture the proteins that support their prodigious reproductive efforts, they remove only a small proportion of the sugar from the plant sap. Therefore the liquid they excrete, honeydew, still contains a high concentration of sugars, making a tasty treat for ants. But there seems to be more to the relationship of cotton aphids and Argentine ants than just an exchange of honeydew for protection. Researchers have found that cotton aphids actually reproduce more rapidly in the presence of Argentine ants, even in the laboratory where enemies were absent. So the ants must be providing more than protection. A study on another ant-aphid relationship found that aphids live longer and have more young when tended by ants, compared with aphids that are not tended. Ants may encourage aphids to feed more, and perhaps they lower aphid stress by protecting them from enemies. Supporting the idea of reduced stress, aphids tended by ants have been found to be less responsive to alarm pheromones, the volatile chemical produced by endangered aphids that warns others. In any event, certain ants seem to encourage higher fitness, measured by reproductive rates, of certain aphids. This is truly a mutually beneficial relationship.

Tending by Argentine ants also has been shown to influence the appearance and size of cotton aphids. Cotton aphids produce more light green individuals when tended by Argentine ants, and dark green individuals when not tended by them. It seems that light green aphids are less susceptible to parasites, and the Argentine ants, while effective at preventing predation, are less effective at preventing attack by parasites. The benefits of ant protection are significant enough that aphids are also known to modify the composition of their honeydew to be more attractive to ants. On the other hand, Argentine ants need the aphids and other honeydew-producing insects that they tend. Argentine ant colony growth and survival is improved by access to cotton aphids. The ants and aphids that engage in mutualistic associations have formed a relationship in which each organism adapts to meet the needs of the other and by doing so meets its own needs. More broadly, this is a relationship that shows that actions that benefit another organism, allowing them to flourish, may result in increased fitness for the organism that initiates the action.

Our human relationship with nature ranges from romantic worship to unbridled greed in the use of natural resources. There are similar examples in nature in which animals deplete a resource to the extent that it is detrimental to the animal itself. Deer starve, for example, when their populations grow as a result of reduced predation, and they consume all available food resources. Plants and the insects that feed on them have a long evolutionary history of adaptations to each other, maintaining a sort of cyclic balance between the two. Plants evolve defensive chemistries, physical defenses, or reproductive strategies that limit or deter feeding damage by herbivores. Biologists often talk about an evolutionary arms race between plants and the organisms that feed on them. The plants evolve a chemical defense, and the herbivore evolves a way of mitigating the toxic effects of the chemical. The adaptation and response between plant and herbivore is a process called coevolution. Coevolution can result in a very specific set of herbivores that feed on a certain plant—those herbivores that have developed a way to deal with that plant's defenses. Solving a plant's defensive puzzle may benefit an herbivore in reducing competition if other herbivores are unable to do so. The benefit in being one of a limited number of successful herbivores on a well-defended plant may be significant enough to result in specialization in feeding on that particular plant species, genus, or family. Milkweeds are a fascinating example of coevolution with a specific group of herbivores, including the larvae of the monarch butterfly (*Danaus plexippus*).

Milkweed defenses against herbivores include hairs on leaves, the production of sticky latex when damaged, and chemicals including cardenolides and cysteine proteases. Cardenolides are steroids that can be toxic to many animals. At lower dosages, they are distasteful and may cause vomiting. The cysteine proteases may be toxic to insects and tiny nematode worms. Only a small community of insect herbivores evolved the adaptations necessary to overcome milkweed defenses and are therefore able to feed on the plant. Monarch butterflies are the best-known insect in this group. They, like others in this community, are milkweed specialists, feeding only on milkweed. This has created a problem for monarch butterfly survival, as milkweed plants are becoming harder to find in the Anthropocene. Modern agricultural herbicides have been used to effectively replace milkweed with corn and other crops. Of course, monarch butterflies face other problems too, such as reductions in overwintering habitat in Mexico and California and herbicide use affecting milkweeds across North American breeding grounds. Western overwintering populations in California have dropped by over a million butterflies in the past twenty years and perhaps face extinction in the future. On the bright side, eastern overwintering populations in Mexico increased in 2019, and conservation organizations are now encouraging homeowners to plant milkweed in their gardens as food plants for monarchs.[1]

Milkweed defenses that limit the number of insect herbivores feeding on the plants are even challenging for those insects adapted to them. Tiny monarch butterfly larvae, or caterpillars, first emerging from the egg have a very low survival rate. Observations of these first-stage caterpillars in the field find that less than 11 percent survive to the second stage. Caterpillars emerging from the egg must first mow down plant hairs before they can gain access to the milkweed leaf surface. Additionally, the caterpillars must circumvent the pressurized latex system in the plant. Milkweed exudes sticky, milky latex when damaged. The latex solidifies in contact with air and can "gum-up" the mouthparts of the caterpillars, eventually killing them if they can't free themselves. So the young caterpillars first chew trenches in the leaf that reduce latex flow to the portions of the leaf where the caterpillar will feed. Those initial bites of the leaf and the resultant latex flow are a moment of significant danger to the newly emerged caterpillar. The caterpillar may be stuck to the leaf by latex, or may be killed by cardenolides present in the latex that they ingest in the process.

Monarch butterfly caterpillars that survive, growing to later stages on milkweed, are brightly colored and very visible, with yellow, black, and white stripes encircling their bodies. The well-known and beloved monarch butterfly adults are also brightly colored, displaying their characteristic deep orange and black wings. In fact, most members of the community of milkweed herbivore specialists are brightly colored and conspicuous. The red milkweed beetle (*Tetraopes tetrophtalmus*) adult is bright red with black spots on its back. The small milkweed bug (*Lygaeus kalmii*) is also bright red with larger zones of black revealed on its back. The swamp milkweed leaf beetle (*Labidomera clivicollis*) has an orange background on its back with black splotches, and a black head. It's as though these insects are advertising their presence on milkweed. And, indeed they are.

Biologists refer to these bright colors as aposematic, or warning coloration. The coloring warns potential predators that these insects are distasteful and possibly dangerous. Beautiful and brightly colored to the human eye, they are a warning to birds looking for a meal. Monarch butterfly caterpillars, and their fellow brightly colored herbivores feeding on milkweed, are able to incorporate the cardenolides produced by milkweed into their own bodies. The same cardenolides are then passed on from the younger stages to the adult stage, which in the case of monarchs doesn't feed on milkweed. These insects contain cardenolides at a level that is at least distasteful to predators, if not toxic. In laboratory experiments, when a blue jay (*Cyanocitta cristata*) ate a monarch butterfly, it vomited minutes after ingestion. The same blue jay subsequently refused to feed on monarch butterflies after that experience. The milkweed plant in a sense both feeds and protects its community of herbivores.

This is not a mutualistic relationship, however. Herbivores have a negative effect on milkweed plants, diminishing their growth and reproduction. And in spite of the benefits to specialist herbivores, milkweed defensive chemistry may slow the growth of the insects that feed on it, and even kill them. The relationship between milkweed and its herbivores has developed over evolutionary time, and continues to be fine-tuned between participants today.

Self-interest is the basis for all of the relationships described in this chapter. All organisms strive to flourish and reproduce in their own way. The route to this goal, though, involves interaction with others—whether members of the same species or other species. Voles are an example of animals that regulate their numbers within a population according to the carrying capacity of their environment. Their evolved behavior, limiting numbers according to population levels, must play a role in their success. Like voles, human beings receive feedback from nature as our resource use exceeds the capacity of our environment. Sometimes it is easy to see, as when overpopulation produces deforestation and erosion. Sometimes our effect on the environment is more difficult to discern, and we require the use of scientific tools, an expansion of our sensing abilities, to identify changes like increases in atmospheric carbon dioxide—a key factor in climate change. Just as voles modify populations in accordance with the capacity of their environment, we humans must also either lower our population growth rate or reduce our resource consumption in order to continue to flourish. Further, as we will discuss in the next chapter, there may be moral reasons to consider the impact of human resource consumption on other species as well as our own. Science and messages from our own senses provide feedback from our environment that suggest self-regulating behavior to insure the success of human and other organisms.

We can find in human life multiple examples of the kinds of mutually beneficial relationships that we observe in ants and acacias, and with ants and aphids. We cultivate a variety of plants that provide fruits for our consumption. We ensure that apple and walnut trees have enough water and space to flourish, producing the reproductive plant parts that we use for food. This kind of relationship seems to most closely resemble that of aphids and ants in which ants tend their aphids. Another way of looking at the ant and acacia relationship is seeing the ant as a defender of another species while deriving benefits from that species. The parallel to human beings seems clear. We too may see ourselves, even more broadly, as defenders of nature all the while accepting myriad benefits. This includes not just domesticated nature, such as apples and walnuts, but wild nature. We can defend the ecosystem that contributes to producing clean, fresh water for our cities or we may act in simply defending wilderness because it serves to revive our spirit. That wildness may intrude in your basement or backyard, and though uninvited and small, it may also revive our spirit. We can rejoice in the ant and aphid, the silverfish, or the rat-tailed maggot.

As you might imagine, the development of pesticide products requires a constant source of the organism you wish to kill, for testing purposes. Insect rearing facilities, or insectaries, can achieve an almost industrial efficiency in turning out cockroaches, ants, flies, and mosquitoes. It requires cleanliness to prevent disease among the animals, and a certain amount of scientific rigor to produce hundreds or even thousands of healthy insects of the same age. All of these animals are sacrificed in the pursuit of faster acting, longer lasting, or just more effective bug killers and repellers. One of the issues for those doing the product testing is how well these insects, the product of many laboratory-reared generations, represent their wild brothers and sisters. Wild cockroaches, living it up in your apartment, may have, over many generations residing between walls and in cabinets, been exposed to a variety of pesticides, developing resistance to many of them. Laboratory roaches may not necessarily have a good life, but it's an easy one. They don't have to go far to find food or shelter. As a result, sometimes products carefully formulated to kill roaches in the lab will fail against the wild roaches that lead a rough-and-tumble life under the sink.

There is something a little sad about the volume of life that passes through the insectary in climate-controlled, artificially lit, and caged surroundings. You might say these creatures live (and die) in service to humankind to make our lives easier. But there is much waste and many are killed unnecessarily in projects oriented not to better lives but to increasing sales. This is a one-sided relationship. These flies will never see a manure pat. The roaches won't navigate someone's plumbing. The mosquitoes will never sink their proboscis into a fleshy arm. These are bits of nature torn from their environment to build a better roach trap.

It may seem peculiar to be concerned about the lives or fate of insects bred for testing pesticides, but I think a lack of empathy is perhaps fundamental to the present-day insect apocalypse— the global loss of large numbers of insects and insect species. When vertebrate animals are used for research, they are afforded a number of safeguards for their care and welfare. Maybe we should do the same for invertebrates. This conversation would at least get us thinking about how we relate to these smaller organisms and what our responsibilities might be.

The relationship between milkweed and its community of herbivorous insects is not one of mutual benefit either. Insects that feed on milkweed are consumers and provide little in the way of return to the plant that we know of. Their ability to metabolically solve the chemical puzzle of milkweed defenses provides several advantages, as we have seen. Because a limited number of insect species can feed on milkweed, competition for this food resource is restricted. And the ability to incorporate milkweed chemical defenses, particularly cardenolides, provides these insects with their own defense system against predators. If there is a lesson for humans here, it could be using what nature gives us without destroying it. This is not to say that the milkweed-feeders are careful about how much milkweed they eat. It is that there is a finely tuned relationship between the consumer and consumed that is mediated by the plant's defenses and reproductive capabilities.

Nature provides us with nutrients, medicines, and raw materials that we use for sustenance, protection from the elements, and construction of tools. Although nature is resilient, our increasingly sophisticated and intense utilization of natural resources now often exceeds the natural capability to regenerate. Fish stocks decline to the point that we can no longer exploit them. Biodiversity decreases around the world, limiting our ability to learn from those species that go extinct. Here is a grand opportunity for human beings to demonstrate that our large brain size and reasoning capacity is a successful evolutionary trait. We can make the requisite adjustments to our use of natural resources that will preserve them for the future.

EIGHT

Valuing Our Anthropocene Partners

A PILLBUG SCUTTLES ALONG the baseboard. A spider has taken up residence and built a web in a corner of the ceiling. Ants are searching for food on the kitchen floor. Bright yellow dandelions are scattered in the front lawn. How do we react to these organisms in and around our home? Killing them is a big business. Marketing materials for products like Raid and Ortho cry out for action. They tell us to "attack bugs," "kill the bugs you don't see," "protect your home from bugs," and "prevent bugs." We should develop a "home defense" that may "create a barrier that will kill bugs you have and prevent new bugs for up to 12 months." We are under siege from an insidious foe!

We can also kill more than 250 kinds of weeds in hours, according to a product label, if we have the time and the inclination. These products are big sellers, as testified by their ubiquity in stores and the money their companies spend on advertising. Why would we want to protect or defend our home from bugs? The bugs described by Raid and Ortho might include any variety of crawling or flying, small, multilegged organisms (arthropods), most of which are no threat to human beings. Similarly, weeds in the lawn generally have no effect on human well-being. Yet we spend quite a bit of time and money trying to eradicate them. I have contributed to killing innumerable insects, given the global reach of the products on which I worked. I have also been a killer in my own kitchen and front yard. What are the emotional and rational underpinnings of our response to nature that embraces us most closely? How might we reconsider these relationships?

I began to think more deeply about how we treat small creatures while studying religion, ethics, and the environment in a two-year program at Yale University. I had retired early from the pesticide company and wanted to pursue broader interests in why we should care for nature. The question of how we value other organisms relative to human beings came up in the middle of my very first semester, in a class involving the intersection of religion, science, and ethics. Do the needs and desires of human beings always come first? How we treat non-human life became a subtext for much of my studies. It was something I was thinking about throughout this introduction to religion and philosophy. In the process, I came to believe that we have ethical responsibilities toward even the smaller parts of the natural world and that our consciousness of the organisms that we live most intimately with is the place our environmental concerns must begin.

A friend tells me that her daughter once enjoyed picking up and examining insects in the garden when she was young, but since attending school the girl's views have changed. She will no longer touch bugs and says they are "icky." This suggests that our attitude toward insects is a product of socialization and that we are learning from others to dislike them. The word "icky" is defined in the *Merriam-Webster Dictionary* as something offensive to the senses or sensibilities—distasteful. Although icky might be a child's word, I often hear the word "disgusting" applied to

insects, spiders, and the like. It seems that many people feel revulsion toward small multilegged creatures. But what is it about
these animals that inspires an urge in humans to kill them?

Perusing the literature on reactions to arthropods, it's clear
the emotions of fear and disgust play a prominent role. Some
psychologists believe that we have inherited a fear of spiders
in particular. Widow spiders (in the genus *Latrodectus*), such as
black widows in North America, were present in Africa when
humans evolved. Widow spider venom is a potent neurotoxin,
and the bite can cause pain, nausea, and other symptoms that,
without treatment, may disable a person for days—though death
is very rare. It is speculated that avoiding extended sickness from
these spider bites would be a beneficial evolutionary adaptation,
leading to the evolved emotion of fear and spider avoidance.
Studies of the fear of spiders often use spider images to elicit
reactions. Research meant to explore the evolved behavior idea
involved showing pictures of spiders, snakes, fish, and flowers to
six-month-old babies. Researchers found that the infants' pupils
dilated in response to spiders and snakes significantly more than
to fish and flowers. They conclude that this evidence of arousal
shows a predisposal to development of a fear response to spiders
and snakes. Children eighteen to thirty-six months old display
more caution toward a tarantula spider in a container than a caged
hamster or fish in a bowl, but interact with the caged tarantula
and do not avoid it. So fear does not appear to be fully developed
at these younger ages. The age at which fear of spiders actually

develops reportedly ranges from four to ten years in various studies.[1] We therefore seem to be predisposed to fear spiders, but development of the fear requires some socialization. Maybe there is an opportunity for a more reasoned approach to spiders and other arthropods to overcome our predisposition to fear them.

Whether it's inherited evolutionary adaptation or learning, or a combination of both, many people fear spiders—even entomologists, including some who surprisingly mention spiders as their worst fear. But fear of spiders does not appear to extend more generally to all arthropods. Identity is important. In a study of college students, spiders caused more fear than bees and wasps, which were more feared in turn than beetles, butterflies, and moths. But it's not just fear. The same study found that spiders are considered more disgusting than the insects used for comparison.[2] Spiders, then, are an extreme case of these negative reactions, having two emotions working against them—fear and disgust. David Quammen describes watching a black widow spider (*Latrodectus hesperus*) and its egg sac on his desk. Eventually the eggs hatched and, confronted with many tiny dispersing black widows, he reached for a can of Raid. Quammen suggests making eye contact with an organism before you decide what to do about it. When he accomplished eye contact with the spider, he says he felt revulsion, fear, and fascination.[3] Of course, eye contact with arthropods can be tricky since they often have many (black widows have eight) to your two.

After spiders, disgust in the study among college students was equally applied to beetles, wasps, and bees, but butterflies and moths were less disgusting. The word "disgust" signifies an oral sensation from its Latin roots, meaning negative taste. So presumably the idea of finding spiders, beetles, and wasps in your mouth is disgusting—less so butterflies. People will not countenance eating a cockroach, even when it's sterilized to eliminate germs. But this can be cultural. Some people are quite happy to eat grasshoppers, such as chapulines in Mexico, though others find them disgusting. Is it just how they look? Is it the crunchy exoskeleton? Or is it what's contained inside the exoskeleton? Chapulines seem to be making cultural inroads, though; they are now sold as ballpark snacks at Seattle Mariners baseball games. Part of our aversion to arthropods may be that we believe they are germ-laden and will sicken us. There is certainly some truth to that concern, because scientists have identified human pathogens on the bodies of insects such as house flies and cockroaches, both of which are often seen in association with garbage. In this respect, you legitimately might not want a fly to land on your birthday cake or a cockroach to crawl across your plate. While some concern is reasonable in these instances, revulsion toward these insects or reviling them is not. We can appreciate them and still take steps to protect our food and utensils.

Other creatures found in the household, such as pillbugs, silverfish, or fruit flies, are not likely to transmit disease-causing microorganisms. But people still think of them as dirty. Some scientists suggest that there may have been another evolutionary adaptive benefit in this sort of disgust, which might be involved in keeping a clean nest area and reducing the danger of disease.[4] Perhaps, but some scientists also believe that we may have gone too far in our phobia about germs, and our attempts to produce a germ-free household actually have negative health consequences. Rob Dunn, mentioned in Chapter 6, would agree. He and his research team at North Carolina State University have identified bacteria present in locations throughout the home, including hot water heaters, shower heads, and countertops. It really isn't possible to create a bacteria-free environment at home, nor is it good for us.

Similarly, I know from talking to people about their experience with insects in homes that many of them feel that the presence of insects signifies a dirty home. Even the tidiest homes, though, host plenty of living things, from bacteria to silverfish to fruit flies. So the presence of a pillbug or two doesn't say anything about your housekeeping. In fact, a study in North Carolina (Rob Dunn's team again) in 2017 found that differences in clutter and dust did not affect arthropod diversity in a home.[5] You could say that our preoccupation with removing weeds from the lawn is comparable. A weed-free lawn is also a sort of clean environment. Many homeowner associations even require a pristine lawn, free from weeds. Weeds are not welcome in these neighborhoods. In that way, rejection of weeds also has an element of socialization. It is true that they interfere with the growth of desired plants by competing for resources. Oftentimes, scrappy weeds outcompete vegetables, ornamental plants, or lawn grasses because the weed is better adapted to that environment than the human-favored plants. It is often said that a weed is simply a plant out of place. But it is out of place from a human perspective, not in nature. I suggest that we acquiesce to nature in our homes and gardens, living as best we can with the bugs and plants that naturally occur there. This might include gardening with native plants and living with the odd pillbug or fruit fly that enters our home.

There is a lot to appreciate about the lesser-known living creatures in our immediate environment and reason to respect them. Beyond appreciation though, there is a line of thinking that we have a moral responsibility to them. This approach is derived from a rational consideration of our relationship to nature that might override the emotional reactions and socialization that we have just considered. Given our cultural norms, it seems odd to think that we should leave a weed in place or ignore a centipede in the basement. But I think we should give this way of thinking serious consideration. What was once a threat is less so today, because we live in a much more controlled environment. In the remainder of this chapter, drawing on philosophical, religious, and scientific thought, I will explore the rational support for the idea that we owe moral consideration to even the smaller and more commonplace forms of life—meaning that we should care about what's good or bad for their interests; our actions are right when they are oriented toward the flourishing of these creatures and wrong when our actions interfere with or prevent their flourishing. We can see our actions as good or bad because we can discern that an organism can flourish or suffer from our action. This results in obligations to objects of our moral consideration or concern.

What does it mean if our moral concerns do not end at the dividing line between humans and other species? Just where that dividing line occurs is the subject of a great deal of philosophical argument. It may not be surprising to learn that ants, spiders, and dandelions are often found on the wrong side of the line. The thinking that these organisms are less worthy is most easily seen to begin with Aristotle, who believed that humans, as rational beings, ranked above all other animals, which were merely instinctual. Aristotle arrived at his conclusions through careful observation of other animals, including their anatomy, behavior, physiology, and habitats. He thought of nature as ranging from least to greatest perfection, which was later elaborated by others as the scala naturae, or great chain of being.[6] In ranking these living things, Aristotle considered temperature, moistness, whether an organism had vegetative characteristics or was sentient, and rationality.[7] Human beings were highest on the list because only they had rationality, in addition to being warm and moist, and thus were most perfect of the living beings. Insects were sentient, but were cold and dry, placing them well down the list. Weeds and other plants would fall even lower, being only vegetative and not sentient.

But even though there is general agreement today that humans have superior cognitive powers to any other animal, we also have come to realize that other animals, including insects, have cognitive abilities that challenge the distinction, including tool making, communication, and social behavior. Furthermore, the development of an understanding of evolution demonstrated that living organisms cannot be classified on a single scale of increasing perfection. Instead, we now represent the range of living beings in a more bushy form, with multiple stems each radiating out in multiple ancestral branches. Do we continue to think in Aristotelian terms of living things ordered in degrees of perfection as they approach the pinnacle represented by humankind, and in this way discount the value of lesser organisms like silverfish, crabgrass, and centipedes? Does it become more permissible to kill an organism as we move down the great chain of being from humans to mammals to birds to insects to plants?

Philosopher René Descartes describes an experiment in which he opened the chest of a live rabbit, removed the ribs to reveal the beating heart, and then tied the throbbing aorta with thread.[8] Vivisection, or the dissection of a live animal, was apparently not an unusual practice in the seventeenth century. But this practice also reflects Descartes's contention that rationality separates humans and animals. Animals for him were simply machines—they could not engage in thought. If they lack thought, and therefore consciousness, Descartes may have reasoned, then they must not be conscious of pain. If animals, including insects, spiders, and the like, are nothing more than machines, then they likely have no moral standing—although, in a sign of how much things have changed, I should add that there are discussions today about our moral obligations to robots. Setting aside the question of the degree to which various animals engage in thought, we might ask whether rationality by itself determines whether an organism is deserving of moral consideration. Given what he knew at the time, Descartes might argue that we cannot act in a right or a wrong way toward an insect, since it lacks reason or consciousness. Science, though, has refuted Descartes's contention that animals lack consciousness. This is certainly true of birds and mammals, and even insects are now suggested to have a level of consciousness involving basic subjective experience. Certainly to Descartes's way of thinking in the absence of our more recent scientific discoveries, though, plants and insects are not deserving of moral consideration. Through the eighteenth century, plants and insects would be seen as machines, of lesser perfection in the great chain of being, and nonrational, thus having no moral value. The ideas Descartes had about animals live on. Even today, Stephen Kellert found in research on values and interests related to wildlife that a majority of Americans believe that invertebrate animals are unable to experience consciousness or pain.[9]

Things began to change for non-human species later in the eighteenth century, when the philosopher Jeremy Bentham famously said, "The question is not, Can they reason? nor, Can they talk? but, Can they suffer?"[10] Bentham believed that animals could suffer, and that their suffering should be part of the calculation of the ethics of an action. After Bentham, we fast-forward to the twentieth century, when philosopher Peter Singer published the book *Animal Liberation*, which maintained that all sentient animals should be included when considering the ethics of an action. Singer, like Bentham, points out that many human beings, such as babies or the elderly, might not meet the requirements of rationality set by Descartes. Should they be treated as we treat animals like chimpanzees or cattle that purportedly lack rationality? In Singer's view, we engage in speciesism, discrimination against other species, in our bias toward *Homo sapiens*, which he finds indefensible.[11] Exactly where the sentient line would be drawn is unclear since there is uncertainty among scientists, but certainly birds and mammals would be included. In Singer's view, the list of beings that are sentient would not include insects, although he admits that recent scientific theorizing about insect consciousness might change things.[12] So there is an expansion in what Singer calls the circle of moral concern, as it enlarges to incorporate many organisms beyond human beings. But the circle has not expanded enough to incorporate insects and plants.

Other philosophers ask whether there should be a difference in human moral regard toward living things at all. They believe that rationality or other ways of separating organisms is arbitrary and thus untenable. In their view, even plants and insects should be worthy of moral consideration. Under these scenarios, we should be concerned about the good of dandelions and silverfish. Albert Schweitzer advocated what he called a cosmic ethic that universalizes altruism. He believed that ethics should not stop at human beings but should extend to all of life. Schweitzer contrasts the starting point of his philosophy with that of Descartes. Descartes famously says that "I think, therefore I am." Schweitzer says, "I am life which wills to live, in the midst of life which wills to live." Schweitzer concludes that "it is good to maintain and to encourage life; it is bad to destroy life or obstruct it." He says that we must live out a reverence for life in all of its forms.[13] Apparently, Schweitzer lived out his philosophy. His obituary mentions that the discovery of ants in his hospital would result in quite a bit of disruption as everyone worked to accommodate them. Mosquitoes weren't swatted there and insecticides weren't used.[14] He maintained that a truly ethical person would be careful not to tear a leaf from a tree, break off a flower, or crush an insect as he or she walks. This is a truly enlightened view, from the perspective of a fruit fly or dandelion. Schweitzer would also close his windows and sweat out a hot tropical evening rather than have insects fly to his lamp and die in the flame.

Later in the twentieth century, philosophers such as Paul Taylor and Kenneth Goodpaster have also argued that even the least of life deserves moral consideration. Like Schweitzer, they advocate an attitude of respect for nature. Taylor recognizes that all living things are directed toward their own ends (like growth and reproduction) and that circumstances that inhibit those ends are bad for the organism. Circumstances that favor the organism's ends are good. Thus, according to Goodpaster, even though a plant or an insect may not be conscious of it, that organism has interests. Just as human ethics requires us to respect the interests of other persons, we should also respect the interests of these living things. Taylor maintains that each individual organism has what he calls inherent worth (other philosophers refer to this as intrinsic value), meaning that ants, centipedes, and dandelions deserve moral consideration in and of themselves. The honey bee is valued not because of its importance to human beings, but solely for what it is.[15]

Taylor's environmental ethic finds equal value in all organisms, including crabgrass and silverfish. This is a difficult concept for us. Most of us would not be inclined to say that the life of a wildflower is of equal value to that of a human being. Taylor does say that humans should not suffer in pursuing his ethic. Human beings must fulfill their basic needs when there is conflict, but may be required to compensate nature—when habitat is destroyed, for example. Taylor opposes the destruction of nature for reasons other than need, such as convenience or efficiency. This creates some interesting questions concerning nature at home. If we were to follow his principles, would we replant dandelions in other parts of the yard when we pull them from the lawn? Would we leave some clutter in the basement during spring cleaning so that silverfish continue to feel at home? Goodpaster has come to similar conclusions about the moral considerability of nature. But, he suggests, although all species may have moral considerability, some have greater moral significance than others. So there may be a hierarchy of value that should be employed in situations of human conflict with other species. Despite the oddity of these ideas, we have seen the circle of moral concern gradually widen over time to incorporate the fullness of human diversity. Perhaps those changes seemed strange at the time, too. Maybe it is time to extend our moral consideration to bacteria, crabgrass, and fruit flies. Based on Goodpaster's view, we need not believe that all species have equal value to have a moral commitment to refrain from hampering their flourishing when we can. Thinking this way, we might not be so quick to pull a weed or to spray insecticides when we encounter silverfish. Such a position can be found in the world's religions, a second source of rational consideration of how we interact with living organisms.

World religions may also have contributed to an attitude that discounts the value of the smaller living things. Historian Lynn White, Jr., advanced the thesis that the application of technology and science leading to our ecological crisis was based on Western Christian worldviews. In particular, he mentions the interpretation of biblical scripture that maintains that the earth and its inhabitants were intended to serve the needs of human beings—who are told they have dominion over it and should subdue it.[16] This would indicate that the plants and animals we have been discussing have little or no value unless they serve as a resource for humans. Some Muslims and Jews have arrived at a similar interpretation of their scriptural traditions. Yet, as we will see, there is much countervailing scripture and understanding in each of these traditions that I think more than overcome the point of view that regards nature as a mere human resource.

A second problem posed by world religions for plants and insects is their human-centered and otherworldly orientation. You might ask whether other organisms matter at all if the sole aim of our existence on the earth is to graduate to another dimension. Some Buddhist and Hindu beliefs question the reality of our perception of these individual organisms. And all of these religions anticipate some sort of better existence after death. These beliefs may have a tendency to disvalue the living things that are part of our present reality here on earth. Despite all of this, each of the world religions engages us in a moral relationship with our present reality. So, although it's possible to weigh otherworldliness and human focus too heavily, there is much in these traditions that guides us in how we might relate to silverfish and dandelions.

Nature may be valued as something created by God in Judaism, Christianity, and Islam. In the Hebrew Bible, God reflects on creation and calls it good—that would be all of creation, including the weeds and insects in alleyways, our kitchens, and backyards. There is an acknowledgment in Islam, Judaism, and Christianity of the value of such small creatures. Spiders, for example, hold a special place in Islam. In the Islamic Hadith, the collected sayings and actions of the Prophet Muhammad, it is said that spiders helped conceal him and his companion in a cave by building a web across the entrance so it appeared unoccupied to their pursuers (Musnad Ahmad ibn Hanbal 3251). Two chapters in the Qur'an are even named after the bees and the ants, respectively. In the Hebrew Bible, God makes use of insects, such as locusts, lice, and hornets, as weapons against the Egyptians and the Canaanites. The Hebrew Bible, as mentioned earlier, encourages us to "go to the ant, you lazybones; consider its ways and be wise" (Proverbs 6:6, NRSV).

In the Qur'an, Prophet Sulaiman, marching with his army, hears an ant calling to its fellows to run back to their homes to avoid being trampled. Prophet Sulaiman, who understands the ant, smiles and, considering it a gift, thanks God for God's favor (Qur'an 27:18,19). Bees are celebrated in the Hebrew Bible because, although small, they produce a wonderful product (Sirach 11:3). In another Hadith, there is a story of the Prophet's companion who crumbled bread for ants, saying, "they are our neighbors and have rights over us."[17] Concern for ants, creatures that have no direct benefit for humans, speaks to a broader concern for nature itself in Islam. Some say that the Hebrew Bible and the Qur'an are written revelation, whereas nature is the physical part of God's revelation. Perhaps living with and knowing something about silverfish, fruit flies, crabgrass, and dandelions tells us a little more about and brings us closer to God.

Compassion is a common practice across religions that can be expressed in caring for nature. It reminds me of a paradigmatic moment in a film, *Pad Yatra*, which depicts a Buddhist trek in the Himalayas to benefit the environment. At one point, we see Buddhist monks on their hands and knees blowing ants off the road to prevent their being trampled by walkers or run over by cars.[18] Buddhist concerns about suffering fuel an orientation toward compassion for all living beings, including ants—and some Buddhists extend their concern toward nonliving entities as well. Cows often wander free in India and may be seen traversing major thoroughfares. Indian cows may be garlanded or fed treats by Hindus. Cows are a symbol of the earth in Hinduism; like the earth, they nourish us and ask nothing in return. Gandhi called the protection of cows the most important manifestation of Hinduism, and believed that the protection of cows symbolizes the protection of all that is helpless and weak in the world—such as hover flies that pollinate and pillbugs and millipedes that nourish the earth and us by decomposing decaying plant materials. Compassion or empathy toward living things is a second principle of importance in guiding our relationship to plants and insects.[19]

Religious traditions may call on followers to care for other beings in an extension of the golden rule. Jewish scholar Hava Tirosh-Samuelson says that at its foundation, Judaism requires adherents to care for others. She suggests that if the covenant is expanded to include the earth, then Jews may be obligated to protect other species.[20] Similarly, for Christians, Jesus invokes the golden rule, saying "do unto others as you would have them do unto you" (Matthew 7:12). Can the other be a fruit fly? It seems that Buddhist monk Thich Nhat Hanh would disagree with the philosophers who limit our moral commitment to living beings based on rationality, sentience, or even life. Instead, he says that the Buddhist believes there is no true separation between life and nonlife. All of life is composed of nonliving materials such as water and minerals. Thich Nhat Hanh believes that we should protect not only living things, but also stones, soil, and the oceans.[21] In Thich Nhat Hanh's Buddhist framework, our small subjects would be deserving of protection.

The principle of nonviolence in many religions is often applied to other species. Although small in numbers, the Jain community merits mention for its members' commitment to their smaller neighbors. Jainist nuns and monks will wear a mask to avoid accidentally inhaling small insects, or gently sweep the path in front of them to brush insects and other crawling animals out of harm's way. Jains eat meals before it gets dark to prevent the possibility of mistakenly eating something that lands on their food. These practices derive from an understanding that harming other organisms, from humans to microbes, has a negative effect on karma. As might be expected, Jainism is a religion of strict vegetarianism. But Jain adherents also avoid harming other organisms in the garments they choose to wear, lifestyle choices, and their professions. A Buddhist text, the Mahavagga, cautions that monks shouldn't intentionally kill a living being—even worms or ants. The Mahavagga also inveighs against the travel of monks during the rainy season, when they might injure tender plant life or kill small beings. Furthermore, it is not permitted to kill ants and spiders in Islam unless they harm you, and then only the one that harmed you, not all of them. Refraining from killing plants and insects that don't represent an imminent threat would be a good way to begin to respect the smaller living things that share our habitats.

Certain religions speak to the connection of human beings to nature, and in that way reflect our understanding of ecology. Buddhism recognizes interconnectedness in the universe. Thich Nhat Hanh says that we are part of a larger self in which, for example, the trees are our lungs.[22] So we are harming ourselves when we harm even smaller beings. In Japanese and Chinese Buddhism, plants and trees have the possibility of spiritual liberation. The potential for liberation means we can say that plants and trees have intrinsic value. The cycle of death and rebirth recognized in Buddhism emphasizes our relatedness to other living things, since the human spirit can inhabit non-human beings, like an ant, in the next life. Hinduism also speaks to the interconnectedness of life. The universe is an expression of underlying being, or Brahman. Emancipation from the cycle of death and rebirth involves, in part, the realization of unity, including unity with dandelions and silverfish. Neo-Confucianist Zhang Zai says in his Western Inscription: "Therefore that which fills the universe I regard as my body and that which directs the universe I consider as my nature. All people are my brothers and sisters, and all things are my companions."[23] We have responsibility toward nature as members of this broader community in Confucianism. It is said that Zhang Zai couldn't bring himself to cut his grass or bushes for just this reason. Realizing our interconnection with other species is an important step toward respectful treatment of the species that occur in your home environment.

Life in one way or another is sacred across religions. Islamic scholar Seyyed Hossein Nasr says Islam understands that the fact that God creates nature in God's wisdom means that all of nature is sacred.[24] I firmly believe that the nature we find in and around our homes and gardens qualifies as sacred nature. Buddhists believe that all life forms have a Buddha nature—the potential for enlightenment. Confucians say that all forms of life contain the principle of heaven. In both the Bible and the Qur'an, it seems that all life forms praise God, giving them a sacred role on earth. In the Islamic Hadith, ants are mentioned as one of the communities that sing praises to God. If so, it would seem wrong to disrupt that community, and indeed, in the Hadith, God reprimands a prophet who destroys an ant nest when he is stung by one of the ants: "Because one ant stung you, you have burned a whole community which glorified me"

(Bukhari 3319). Ants are evidence of God's care for even the smallest creatures, as referenced in the Jewish Babylonian Talmud (Hullin 63a). John Grim, who studies indigenous traditions and ecology, relates that the inner life force of the Maori people is thought to depend on "ethical relations" with other living organisms.[25] Many indigenous traditions find nature inhabited by spirits, and indigenous peoples engage in respectful treatment of the organisms that sustain them.

Despite a focus on the behavior and fate of human beings in religion, humans are also viewed as a part of a greater whole. In Judaism, Christianity, and Islam the world belongs to God, not human beings. Human dominion, to the extent it exists, is rooted in responsibility toward nature. This is often referred to as stewardship in Judaism and Christianity, and vice-regency in Islam. We should remember that this responsibility does not just start beyond the city limits, but we have responsibilities to nature in and around our homes. Humans should also exhibit humility. In Judaism, the Babylonian Talmud (Sanhedrin 38a) says that Adam was created last, so "if a man's mind becomes (too) proud he may be reminded that the gnats preceded him in the order of creation." There are plenty of examples of biblical scripture in which humankind plays a more humble role. Speaking from a whirlwind in the Book of Job, God admonishes Job and his companions, reminding them how little they know of the world and how little they can control (Job 38–41). God asks Job (Job 39: 26,27), "Is it by your wisdom that the hawk soars," and "Is it at your command that the eagle mounts up and makes its nest on high?" Job responds, "I am of small account." In Job, humankind is revealed not as ruler but as humble participant in life. Many indigenous traditions locate humans and other beings as members of a community birthed and nourished by Mother Earth. In another part of the Western Inscription, Neo-Confucianist Zhang Zai says, "Heaven is my father and Earth is my mother, and even such a small creature as I finds an intimate place in their midst." Here, Zhang Zai places humans in humble relationship to the broader universe. We are given many opportunities to express that humility right here at home when confronted by silverfish, house centipedes, or crabgrass.

From these examples, I think it's easy to see that smaller living beings are valued in the world's religions. Not only valued, but also they can be thought of as sacred. Furthermore, we find virtues in the world religions that guide our interactions with these organisms in the direction of humility, compassion, and nonviolence—put away the insecticide and weed killer. The world's religions tell us that we are a part of a larger, interconnected whole, which must include house centipedes, dust mites, and crabgrass. And we have responsibilities toward nature as stewards or vice-regents. There are certainly other principles that we could derive from the world's religions for how we ought to interact with more homely nature, but those articulated here are more than sufficient. Clearly, world religions value organisms like plants and insects and provide principles that require care for nature—not just sentient nature, but perhaps even nonliving nature.

Whether or not we rethink what is meant by dominion and subdue in the Hebrew Bible, there is a scriptural basis in the Judeo-Christian tradition that leads us to understand that non-human organisms are deserving of moral consideration. Perhaps Lynn White was right about the influence of the Judeo-Christian tradition on Western culture and resultant environmental degradation. But I think we have come to understand that this is a very narrow reading of the scriptural tradition and that there are other, broader influences in the scripture of both Judaism and Christianity—influences that lead us to humility and to a moral concern for other life on the earth, our neighbors.

I have touched on the role that the sciences play in contributing to how we conceive of our relationship with nature. The biological sciences in particular, over the past two hundred years, augment our understanding of this link to nature, emphasizing human relatedness to the rest of life. We have a deep connection to all of life—genetically, evolutionarily, and ecologically. Each individual organism expresses our genetic and evolutionary connection in the portions of the genome that it shares with us. Like it or not, we are related to and interdependent with silverfish, dandelions, fruit flies, and crabgrass. Aldo Leopold describes this relationship as a kinship that we have with fellow evolutionary travelers. Environmental philosopher J. Baird Callicott elaborates on Leopold's concept of kinship with other organisms, saying, "We should feel and thus behave toward other living things in ways similar to the way we feel and behave toward our human kin."[26] But it seems that our relationship to these organisms today still reflects earlier positions on the limited value of nonhuman life, in which human relatedness to life was not well understood, and non-human life was regarded solely as a resource for human use and machine-like other—particularly those living things at the lower end of the great chain of being. As a result, many give little thought to poisoning, pulling, or crushing the plants and animals we deem pests.

Since Darwin's discoveries, we have begun to appreciate the significance of fellow organisms that have developed through the crucible of selection and evolution—such as the silverfish, surviving over hundreds of millions of years. The combined metabolic, physical, and behavioral adaptations developed by silverfish have demonstrated survival value over that time. Each individual silverfish carries a genome that is the product of a long process of selection and evolution. Another environmental philosopher, Holmes Rolston III, suggests that the value of a living thing comes from the traits produced by its evolved genome. He describes the design of the dragonfly wing, immunology in vertebrate animals, and the internal clock of blue-green algae as traits that have value to the species that carry them. These traits were of value to these organisms long before humans arrived.[27] They allow them to persist as individuals today and as species over evolutionary time—something that can be appreciated and lead to respect. Rolston highlights the marvelous traits of two smaller living things in his examples.

The role of each species and its individuals in the ecosystem is largely described by its ecological niche and place on the food chain. The niche of a species characterizes how it relates to other organisms and its environment and defines the conditions necessary for flourishing—like a silverfish living in a damp basement where it finds food and suitable habitat. An organism's role in the ecosystem is another kind of biological value it carries. Callicott goes further to say that ecological theory unveils "social integration of human and nonhuman nature."[28] The role of many species in their ecosystem was also established before human beings evolved. This biological value becomes apparent when we consider the ecosystem effect of removal or extinction of an organism, such as the possible effect of the loss of larger predators on the American burying beetle, and the prospect of a loss of honey bees and other pollinators. The loss of a species in an ecosystem can have cascading effects. Furthermore, we humans are in reality an assemblage of organisms ourselves—our own ecosystem. We carry six pounds of associated bacteria, our microbiome, affecting how we digest our food and even our emotional state—not to mention an assortment of other organisms at home on the exterior of our bodies. You might argue that we are not ourselves without these microorganisms. When we accord moral concern to other persons, then their microorganisms must be a part of what we value—we are so intimately connected that we cannot separate one from the other.

The recognition of biological value accruing to living things is a product of advances in the study of genetics, evolution, and ecology since the nineteenth century. Earlier philosophers and theologians did not have the benefit of this knowledge. I think that the biological conceptions of value (both evolutionary and ecological) at least lead to appreciation and respect for other species. And I believe it enhances the conception of intrinsic value as part of what makes an organism worthy of moral consideration. Philosophers like Taylor and Goodpaster consider the combination of value and interests—at minimum an interest in survival and reproduction, described by Schweitzer as will to live—as enough to cause us to concern ourselves with what is good or bad for each living thing.

There is a confluence of thought among certain strands of philosophy, religion, and the sciences, leading to an understanding that actions that harm the tiny living things we're considering here may be wrong and that those that promote their good are right. Of course, it's not quite as simple as that. Human beings like all other animals must eat, construct homes, and protect themselves from other organisms and their environment. We know that other beings will be harmed in the process. Most philosophers and religious practitioners address this kind of conflict by either constructing hierarchies of value for organisms, in which humans have priority, or developing principles that determine when it is permissible to interfere with another organism's flourishing. All of the religious traditions we discussed are to one degree or another anthropocentric, or human centered. So in these traditions human needs take precedence over those of other organisms. Yet these religions consistently also encourage us to interact with the other with humility and compassion, and to limit our greed. Most environmental philosophies also value humans over other organisms, but would agree that we have a responsibility toward other species. Therefore, science and sources of wisdom and reason point us in the direction of concern and responsibility for fruit flies and crabgrass.

I have in this final chapter identified the emotions of fear and disgust, our socialization, a view of animals as machines, a hierarchical great chain of being, and an understanding of human as meant to dominate nature as underlying factors in our often hardened approach to nature that surrounds us. Many people also worry that our increasingly urban populace, living in human-created and human-dominated environments, is becoming alienated from nature. This alienation is thought to lead to the desecration of the environment. Since nature is not part of our daily existence, we no longer value it—or so the thinking goes. In reality, though, as we have seen in previous chapters, despite our best efforts, nature interpenetrates our human-created environment. The problem is not that we aren't exposed to nature but that we ignore it or find it to be an inconvenience. Not only does life spring up in vacant lots, in cracks in the sidewalk, and in our pantries, but we also assign nature a place in our towns and cities. We landscape streets with trees and bushes. We plan parks and green space in urban areas. I suspect that the bigger problem is not exposure to nature but how we experience and act toward the nature we encounter. I fear that we relate to nature around us in the way that Descartes seems to suggest—seeing other organisms as unfeeling things or machines rather than living beings.

Our lives today are filled with animated machines from cars to airplanes to the videos you see on your cell phone. Perhaps we are also desensitized to nature by the presence of so much apparent life in nonliving objects. This may be, but more important, we are hardened toward certain non-human life by our feelings, socialization, and how we conceive of other species— our emotions and rational thought. This hardness allows us to take life without regret. It is a cultural assumption that we will kill the dandelion in our yard, step on the spider in the basement, or spray the ants on the patio with insecticide without compunction. To do otherwise may invite ridicule.

I think it's possible to shift these thoughts and emotions through developing an appreciation for the smaller beings that are common in our lives. As we have seen, dandelions, spiders, and ants are not just things or machines. They need not be feared and aren't really disgusting once you get to know them. There is no need to dominate or subdue them. Each plant or insect has a long evolutionary history, and a unique and fascinating way of living. Learning more about these organisms fosters an appreciation for them. And, perhaps our alienation is not so complete as some fear. We do appreciate nature—at least the nature that stays put in our yards and local parks, and the nature that we can visit in forests, deserts, and on the coasts. We enjoy stately maples on the boulevard in midwestern and northeastern cities in the United States; the beautiful rose bush in the Rose Gardens in Portland; the thrill of seeing a mountain goat in the distance at Glacier National Park. Our appreciation of nature in this regard is an aesthetic and emotional appreciation. We may not know the habits of the mountain goat but enjoy seeing one. We may not know the evolutionary and horticultural history of the rose and yet appreciate the fragrance, color, and form of its flower. What we appreciate in this way, we want to maintain. We want to preserve the habitat that supports the mountain goat. We water our shade trees when it's dry so that they will continue their healthy growth. But our appreciation may be fickle. We may not value those bits of nature that seem out of place in our environment. We should explore how we might begin to appreciate nature more fully, wherever it pops up.

An appreciation for nature leads to respect, but respect may also arise from philosophical or religious commitments. We have seen that there are approaches in philosophy and in world religions that lead to respect for even the smallest of organisms. Beyond human reason and revelation though, I have emphasized that individual animals have a biological value outside that which humans accord them. That value is increasingly better understood by what is revealed to us through scientific endeavors. What we appreciate about nature reflects some of that knowledge. But the value is there whether we appreciate it or not. Previous chapters have emphasized the millions of years of evolutionary history bound up in the genome of every organism on the planet. In the case of silverfish, it is about 400 million years. Every silverfish carries the imprint of that history as it traverses countertops and forages in our kitchen. The evolution of termites has resulted in multiple intricately designed body plans, each of which plays a unique role in the colony. Evolution has produced behavior in termites that enables them to farm fungus, and to build elaborate climate-controlled structures. We can admire the adaptations of the dandelion that make it so successful in disturbed environments. It is not, however, our admiration that makes these traits and evolutionary history valuable. Evolutionary success imparts a value that is not dependent on human assessment. Each organism found in and around your house is in its own way a sort of finely crafted jewel created over hundreds of thousands or millions of years through the heat and pressure of selection. Surely this knowledge enriches our appreciation and respect for these organisms.

Both Paul Taylor and Albert Schweitzer base their ethics on

an attitude of respect, though Schweitzer calls it reverence, for life. In their philosophies, respect for life is an attitude that demands moral consideration for all living things. We have also seen that world religions guide us in the direction of respect for life. The golden rule, treating the other as you would wish to be treated, is found consistently in world religions and in philosophies. The religious and philosophical sources and what we have learned from science would have us expand our consideration of the other to all life. So the little fruit fly becomes our neighbor in a Christian formulation of the golden rule.

Life itself is the happy coincidence of factors that may be difficult to duplicate. We are training our most powerful telescopes on the universe, searching for signs of life. Life on earth exists only because of our planet's distance from the right sort of sun, the impact of meteorites on the earth that carried water and organic chemicals important to life, the size of our planet that generates the right amount of gravity, and the formation of an atmosphere suitable for life. In a way this is similar to the evolutionary success of species. The fact that life occurs through such a remarkable convergence of physical features and circumstances encourages further appreciation for all forms of life. Any living organism is the amazing outgrowth of a process whose probability is statistically infinitesimal and may have occurred only once in the universe. Is this not one more reason for human beings to respect silverfish and crabgrass? Kellert suggests that respect can be generated toward all invertebrates by emphasizing the moral equivalence between more highly regarded members, such as butterflies and honey bees, and charismatic wildlife such as wolves and whales.[29] I say we could accomplish even more if we could simply begin to value the spiders, centipedes, and flies that we encounter every day inside our homes.

I believe that a moral consideration for living things and their relationships is the fundamental basis for human relationship to the environment. Yet what does this mean in real terms? How can we act on our moral concern? What is required of us? Part of the answer as to how we approach our responsibility to the environment is determined by our philosophical and religious commitments. I have endeavored to show that moral concern for living things is called for in credible strands of thought in the biological sciences, philosophy, and religion. Some of the ways of thinking about humans and their relation to living things place humans above other animals when conflicts arise. Other approaches see non-human organisms as having equal value to humans. The end result is often similar, since the equal value approach generally makes allowances that enable humans to use nature to fulfill human needs. I am not too troubled by either approach as long as we maintain an attitude of humility toward and respect for non-human life, and acknowledge that we have moral responsibility toward it. The small organisms we've been discussing strive to survive, grow, and reproduce just as we do. We share a common ancestry that is reflected in our DNA. And, we are connected with these organisms in our ecosystem.

I think the critical question for us becomes how we separate human needs from convenience and greed. An attitude of respect and moral concern for other living organisms should generate a consciousness of the needs of other organisms and how our human activities might impact them. It is especially so in the Anthropocene, when our powers to change the earth are so enormous. This may involve lifestyle choices, choices we make in our work, and decisions we make when we are in direct conflict with an organism. Those choices should reflect respect for other organisms, including silverfish and dandelions. We use pesticides far too easily and far too broadly. Habitat for these organisms is destroyed without sufficient consideration for the affected organisms. I don't have a proposal for new laws or policy to address industrial agriculture or habitat loss. What I suggest is more prosaic—simply changing hearts and minds so that we begin to care for what is good and bad for the smaller living beings in our lives. It seems right to repeat the often-used phrase that we should try to live simply, and to live lightly on the earth. The wisdom of ancient religious traditions provides guidance for how we might do that. I think that could be summarized as accepting responsibility for nature surrounding us with humility and compassion. Living in accord with these principles and greater consciousness of smaller beings would go a long way toward ending the insect apocalypse.

I have organized this book around living things that often don't get much love. Part of the purpose of the tour through human relationships with these organisms is to foster an appreciation for the lives of the most mundane and occasionally annoying beings—to raise awareness and soften attitudes toward them. No matter how we might feel about them, I believe these organisms deserve our respect and moral consideration. This will not immediately stop the rush of plant and arthropod extinctions, but it seems like a good place to start. If we can begin to meet the nature that accompanies us in the Anthropocene with respect, then I think we have a solid basis for the care of the broader environment. If not, and we value only distant nature, then there is something morally inconsistent in our behavior toward nature. We are surrounded every day by wondrous examples of explosive life that fills every nook and cranny of human-created buildings and landscapes. The life that faces us at every turn, no matter how small or seemingly insignificant, deserves treatment worthy of a representative of the marvelous and unlikely development of life on earth. Look again in the basement, where two silverfish meet, shyly touching antennae.

Notes

Introduction

1. *The Creature from the Black Lagoon*, directed by Jack Arnold (1954); *The Shape of Water*, directed by Guillermo del Toro (2017).
2. Leopold, *Sand County Almanac*, 174.

CHAPTER ONE
Anthropocene Winners

1. Mansourian et al., "Wild African *Drosophila*," 3965–3966.
2. Aristotle, *Parts of Animals*, Book 1, Part 5.

CHAPTER TWO
Nature at Work

1. Gilbert, "Imperfect Mimicry," 281.
2. González et al., "Decay of Litter," 37.
3. Shear, "Chemical Defenses," 104; Girardin and Steveson, "Millipedes," 108–109.
4. Shear, "Chemical Defenses," 96.

Notes

CHAPTER THREE
Inadvertent Domestication

1. Acorn, "Dinosaurs Failed to Invent Clothes Moths," 192.
2. Mallis, *Pest Control*, 328.
3. Online Etymology Dictionary, Bug, https://www.etymonline.com/word/bug.
4. Edgecombe and Giribet, "Biology of Centipedes," 158.
5. Cooke, "Hypersensitiveness," 156.
6. Clarke et al., "Child Car Seats," 19.

CHAPTER FOUR
Anthropocene Invasions

1. The Lost Ladybug Project, www.lostladybug.org/about.php.
2. "Canada Thistles," *Prairie Farmer* 16, no. 11 (September 16, 1865): 207; "Canada Thistle," *Indiana Farmer's Guide* (Huntington, Ind.) 33, no. 50 (December 10, 1921): 11; "Canada Thistles," *Prairie Farmer* 7, no. 9 (September 1847): 278.
3. "Canada Thistle," *Plough Boy* (Albany, N.Y.) 1, no. 3 (June 19, 1819): 22.
4. Global Invasive Species Database, Invasive Species Specialist Group, www.iucngisd.org/gisd/species.php?sc=109.
5. Bargielowski et al., "Effects of Interspecific Courtship," *Annals of the Entomological Society of America* 108, no. 4 (2015): 513–518.
6. Worobey et al., "Outdoor Physical Activity," *Journal of the American Mosquito Control Association* 29, no. 1 (2013): 78–80.
7. Krulwich, Robert, "Nice Things About Mosquitoes," *Krulwich Wonders: Krulwich on Science* (2008); https://www.npr.org/templates/story/story.php?storyId=93049810.
8. Six Nations, "The Haudenosaunee Address to the Western World," 85.

CHAPTER FIVE
The Unlucky: Anthropocene Extinctions

1. Furness and Soluk, "Diversion Structures," 450.
2. Quotations and story as related by Daniel Gluesenkamp (Executive Director, California Native Plant Society) are from a telephone interview, May 30, 2018.
3. Quotations from Wyatt Hoback (Assistant Professor, Oklahoma State University) are from a telephone interview, July 26, 2017.

4. U.S. Fish and Wildlife Service, "Reclassifying the American Burying Beetle," 4.
5. Tennyson, Alfred, *In Memoriam* (London: The Bankside Press, 1900), 60.
6. U.S. Congress, *Endangered Species Act of 1973*, Sec. 2(a)(1).

CHAPTER SIX
Human Exceptionalism?

1. Pringle et al., "Spatial Pattern," 5.
2. Hölldobler and Wilson, *The Superorganism*, 4.
3. Franks et al., "House-Hunting Social Insects," 1576.
4. NIH Human Microbiome Project, https://hmpdacc.org/hmp/overview/; Nowakowski et al., "Microbiome and Psychopathology," 69.
5. Knell and Simmons, "Mating Tactics," 2350.
6. Dunn, *Never Home Alone*, 256.
7. Berenbaum, *Maggots, Mites, and Munchers*, xvii.
8. Silverman and Selbach, "Feeding Behavior," 93–102.
9. Barron and Klein, "Origins of Consciousness," 4905.

CHAPTER SEVEN
Coexistence

1. Curry and Kimbrell, "Monarch Butterfly," Center for Biological Diversity, 2019.

CHAPTER EIGHT
Valuing Our Anthropocene Partners

1. New, "Cocktail Party," 167; Hoehl, "Infants," 6; LoBue, "Live Animals," 67; Vetter, "Arachnophobic," 174.
2. Vetter, "Arachnophobic," 170; Gerdes, "Spiders Are Special," 70.
3. Quammen, *Flight of the Iguana*, 3–8.
4. Gerdes, "Spiders Are Special," 70; Rozin, "Disgust," 25, 33.
5. Leong, "Habitats," 5.
6. Lovejoy, "Great Chain of Being," 58.
7. Aristotle, *Complete Works*, 1135–1137.
8. Descartes, *Philosophical Writings*, 81.
9. Kellert, *Value of Life*, 124.
10. Bentham, *Principles of Morals*, footnote 144.

11. Singer, *Animal Liberation*, 6.

12. Singer, "Are Insects Conscious?"

13. Schweitzer, *Civilization*, 242.

14. Reuters, "Schweitzer Dies at His Hospital."

15. Taylor, "Ethics of Respect," 201; Goodpaster, "Morally Considerable," 320.

16. White, "Historical Roots," 1206.

17. Johnson-Davies, "Island of Animals," xvii.

18. *Pad Yatra: A Green Odyssey*, directed by Wendy J. N. Lee (2013).

19. Gandhi, *Hindu Dharma*, 118.

20. Tirosh-Samuelson, "Sources of Judaism," 118.

21. Nhat Hanh, "Sun My Heart," 167.

22. Nhat Hanh, *No Death*, 146.

23. de Bary, *Chinese Tradition*, 683.

24. Nasr, "Contemporary Islamic World," 96.

25. Grim, "Indigenous Traditions," 2618.

26. Leopold, *Sand County*, 109; Callicott, *Land Ethic*, 125.

27. Rolston, "Naturalizing Values," 77–78.

28. Callicott, *Land Ethic*, 82.

29. Kellert, *Value of Life*, 129.

Bibliography

Acorn, John. "How Dinosaurs Failed to Invent Clothes Moths, Among Other Things." *American Entomologist* 54, no. 3 (2008): 191–192.

Adams, Mark D., Susan E. Celniker, Robert A. Holt, Cheryl A. Evans, Jeannine D. Gocayne, Peter G. Amanatides, . . . and J. Craig Venter. "The Genome Sequence of *Drosophila melanogaster*." *Science* 287, no. 5461 (2000): 2185–2195.

Agrawal, Anurag A. "Natural Selection on Common Milkweed (*Asclepias syriaca*) by a Community of Specialized Insect Herbivores." *Evolutionary Ecology Research* 7 (2005): 651–667.

Agrawal, Anurag A., Marc J. Lajeunesse, Mark Fishbein. "Evolution of Latex and Its Constituent Defensive Chemistry in Milkweeds (*Asclepias*): A Phylogenetic Test of Plant Defense Escalation." *Entomologia Experimentalis et Applicata* 128 (2008): 126–138.

Als, Thomas D., Roger Vila, Nikolai P. Kandul, David R. Nash, Shen-Horn Yen, Yu-Feng Hsu, Andre A. Mignault, Jacobus J. Boomsma, and Naomi E. Pierce. "The Evolution of Alternative Parasitic Life Histories in Large Blue Butterflies." *Nature* 432, no. 7015 (2004): 386–390.

Altincicek, Boran, and Andreas Vilcinskas. "Analysis of the Immune-Inducible Transcriptome from Microbial Stress Resistant, Rat-Tailed Maggots of the Drone Fly *Eristalis tenax*." *BMC Genomics* 8 (2007): 326. DOI: 10.1186 /1471-2164-8-326.

Amaral, Michael, Andrea Kozol, and Thomas French. "Conservation Status and Reintroduction of the Endangered American Burying Beetle." *Northeastern Naturalist* 4, no. 3 (1997): 121–132.

Amsalem, Etya, Christina M. Grozinger, Mario Padilla, and Abraham Hefetz. "Chapter Two—The Physiological and Genomic Bases of Bumble Bee Social Behaviour." *Advances in Insect Physiology* 48 (2015): 37–93.

Anderson, Robert S. "On the Decreasing Abundance of *Nicrophorus americanus* Olivier (Coleoptera: Silphidae) in Eastern North America." *Coleopterists Bulletin* 36, no. 2 (1982): 362–365.

Anderson, Robert S., and Stewart B. Peck. "Geographic Patterns of Colour Variation in North American *Nicrophorus* burying beetles (Coleoptera: Silphidae)." *Journal of Natural History* 20, no. 2 (1986): 283–297.

Arlian, Larry G. "Biology, Ecology, and Prevalence of Dust Mites." *Immunology Allergy Clinics of North America* 23 (2003): 443–468.

Aristotle. *On the Parts of Animals*. Translated by William Ogle. (The Internet Classics Archive, 1994–2009) Book 1, Part 5; http://classics.mit.edu /Aristotle/parts_animals.1.i.html.

Aristotle. *Complete Works of Aristotle: The Revised Oxford Translation*. Edited by J. Barnes. Princeton: Princeton University Press, 1984.

Augustine. "The Literal Meaning of Genesis." In *On Genesis*, translated by E. Hill, edited by J. E. Rotelle, 155–506. Hyde Park, N.Y.: New City Press, 2002.

Ayasse, Manfred, and Stefan Jarau. "Chemical Ecology of Bumble Bees." *Annual Review of Entomology* 59 (2014): 299–319.

Bargielowski, Irka, Erik Blosser, and L. Phillip Lounibos. "The Effects of Interspecific Courtship on the Mating Success of *Aedes aegypti* and *Aedes albopictus* (Diptera: Culicidae) Males." *Annals of the Entomological Society of America* 108, no. 4 (2015): 513–518.

Barron, Andrew B., and Colin Klein. "What Insects Can Tell Us About the Origins of Consciousness." *Proceedings of the National Academy of Sciences* 113, no. 18 (2016): 4900–4908.

Bates, Adam J., Jon P. Sadler, Alison J. Fairbrass, Steven J. Falk, James D. Hale, and Tom J. Matthews. "Changing Bee and Hoverfly Pollinator Assemblages Along an Urban-Rural Gradient." *PLoS One* 6, no. 8 (2011): e23459; Doi:10.1371/journal.pone, 0023459, http://journals.plos.org /plosone/article?id=10.1371/journal.pone.0023459.

Bentham, Jeremy. *Introduction to the Principles of Morals and Legislation*. Oxford: Clarendon Press, 1789.

Berenbaum, May R. *Ninety-nine More Maggots, Mites, and Munchers.* Urbana: University of Illinois Press, 1993.

Betz, Robert F., William R. Rommel, and Joseph J. Dichtl. "Insect Herbivores of 12 Milkweed (Asclepias) Species." In *Proceedings of the Fifteenth North American Prairie Conference,* edited by Charles Warwick, 7–19. Bend, Ore.: Natural Areas Association, 1997; http://digital.library.wisc.edu/1711.dl /EcoNatRes.NAPC15.

Blackmore, Lorna M., and Dave Goulson. "Evaluating the Effectiveness of Wildflower Seed Mixes for Boosting Floral Diversity and Bumblebee and Hoverfly Abundance in Urban Areas." *Insect Conservation and Diversity* 7 (2014): 480–484.

Boykin, Laura M., Michael C. Vasey, V. Thomas Parker, and Robert Patterson. "Two Lineages of *Arctostaphylos* (Ericaceae) Identified Using the Internal Transcribed Spacer (ITS) of the Nuclear Genome." *Madrono* 52, no. 3 (2005): 139–147.

Byrne, Marcus, Marie Dacke, Peter Nordstrom, Clarke Scholtz, and Eric Warrant. "Visual Cues Used by Ball-Rolling Dung Beetles for Orientation." *Journal of Comparative Physiology A* 189, no. 6 (2003): 411–418.

CABI Invasive Species Compendium. "*Digitaria sanguinalis* (Large Crabgrass)"; http://www.cabi.org/isc/datasheet/18916.

California Department of Transportation, California Department of Fish and Game, the Presidio Trust, National Park Service, U.S. Fish and Wildlife Service, December 21, 2009. "Memorandum of Agreement Regarding Planning, Development, and Implementation of the Conservation Plan for Franciscan Manzanita (Arctostaphylos franciscana)"; https://cdn.cnsnews.com/documents/MOA%20-%20Fran%20Man %20-%202009.pdf.

Callicott, J. Baird. *In Defense of the Land Ethic: Essays in Environmental Philosophy.* Albany: State University of New York Press, 1989.

Cameron, Sydney A., Jeffrey D. Lozier, James P. Strange, Jonathan B. Koch, Nils Cordes, Leellen F. Solter, and Terry L. Griswold. "Patterns of Widespread Decline in North American Bumble Bees." *Proceedings of the National Academy of Sciences* 108, no. 2 (2011): 662–667.

Cameron, Sydney A., Heather M. Hines, and Paul H. Williams. "A Comprehensive Phylogeny of the Bumble Bees (*Bombus*)." *Biological Journal of the Linnean Society* 91 (2007): 161–188.

"Canada Thistle." *Indiana Farmer's Guide* (Huntington, Ind.) 33, no. 50 (December 10, 1921).

"Canada Thistles." *Prairie Farmer* 7, no. 9 (September 1847).

"Canada Thistles." *Prairie Farmer* 16, no. 11 (September 16, 1865).

Chiras, Daniel D. *Environmental Science* (9th ed.). Burlington, Mass.: Jones & Bartlett Learning, 2013.

Clarke, David, Daniel Burke, Michael Gormally, and Miriam Byrne. "Dynamics of Dust Mite Transfer in Modern Clothing Fabrics." *Annals of Allergy, Asthma and Immunology* 114 (2015): 335–340.

Clarke, David, Michael Gromally, Jerome Sheahan, and Miriam Byrne. "Child Car Seats—A Habitat for House Dust Mites and Reservoir for Harmful Allergens." *Annals of Agriculture and Environmental Medicine* 22, no. 1 (2015): 17–22.

Collett, Thomas S., and Michael F. Land. "Visual Spatial Memory in a Hoverfly." *Journal of Comparative Physiology* 100 (1975): 59–84.

Cooke, Robert A. "Studies in Specific Hypersensitiveness: IV. New Etiologic Factors in Bronchial Asthma." *Journal of Immunology* 7 (1922): 147–162.

Cox, Patrick D., and David B. Pinniger. "Biology, Behavior and Environmentally Sustainable Control of *Tineola bisselliella* (Hummel) (Lepidoptera: Tineidae)." *Journal of Stored Products Research* 43 (2007): 2–32.

Cripps, Michael G., Andre Gassmann, Simon V. Fowler, Graeme W. Bourdot, Alec S. McClay, and Grant R. Edwards. "Classical Biological Control of *Cirsium arvense:* Lessons from the Past." *Biological Control* 57 (2011): 165–174.

Cudney, David W., and Clyde L. Elmore. "Pest Notes: Dandelions." Davis, Calif.: *University of California Agriculture and Natural Resources Publication 7469*, 2006.

Curry, Tierra, and George Kimbrell. "Eastern Monarch Butterfly Population Rebounds: Ideal Weather Conditions Last Year Renew Hope for Beleaguered Butterfly." *Center for Biological Diversity*, Press Release, January 30, 2019; https://www.biologicaldiversity.org/news/press_releases/2019/monarch-butterfly-01-30-2019.php.

Davis, Mark A., Abby Colehour, Jo Daney, Elizabeth Foster, Clare Macmillen, Emily Merrill, Joseph O'Neil, Margaret Pearson, Megan Whitney, Michael D. Anderson, and Jerald J. Dosch. "The Population Dynamics and Ecological Effects of Garlic Mustard, *Alliaria petiolata*, in a Minnesota Oak Woodland." *American Midland Naturalist* 168 (2012): 364–374.

Davis, L. R. "Notes on Beetle Distributions with a Discussion of *Nicrophorus americanus* Olivier and Its Abundance in Collections." *Coleopterists Bulletin* 34, no. 2 (1980): 245–252.

Davis, Mark A., Clare MacMillen, Marta LeFevre-Levy, Casey Dallavalle, Nolan Kriegel, Stephen Tyndel, Yuris Martinez, Michael D. Anderson, and Jerald J. Dosch. "Population and Plant Community Dynamics Involving Garlic Mustard (*Alliaria petiolata*) in a Minnesota Oak Woodland: A Four Year Study." *The Journal of the Torrey Botanical Society* 141, no. 3 (2014): 205–216.

Deans, A. M., Sandy M. Smith, Jay R. Malcolm, William J. Crins, and M. Isabel Bellocq. "Hoverfly (Syrphidae) Communities Respond to Varying Structural Retention After Harvesting in Canadian Peatland Black Spruce Forests." *Environmental Entomology* 36, no. 2 (2007): 308–318.

de Bary, Wm. Theodore, and Irene Bloom. *Sources of Chinese Tradition: From Earliest Times to 1600,* 2nd edition, volume 1. New York: Columbia University Press, 1999.

DeLeon-Rodriguez, Natasha, Terry L. Lathem, Luis M. Rodriguez-R, James M. Barazesh, Bruce E. Anderson, Andreas J. Beyersdorf, Luke D. Ziemba, Michael Bergin, Athanasios Nenes, and Konstantinos T. Konstantinidis. "Microbiome of the Upper Troposphere: Species Composition and Prevalence, Effects of Tropical Storms, and Atmospheric Implications." *Proceedings of the National Academy of Sciences* 110, no. 7 (2013): 2575–2580.

Descartes, René. *The Method, Meditations and Philosophy of Descartes.* Translated by John Veitch. Washington D.C.: M.W. Dunne, 1901.

Descartes, René. *The Philosophical Writings of Descartes, Volume 3: The Correspondence.* Translated by John Cottingham, Dugald Murdoch, Robert Stoothoff, and Anthony Kenny. Cambridge: Cambridge University Press, 1991.

Devineni, Anita V., and Ulrike Heberlein. "The Evolution of *Drosophila melanogaster* as a Model for Alcohol Research." *Annual Review of Neuroscience* 36 (2013): 121–138.

Dinan, Timothy G., Roman M. Stilling, Catherine Stanton, and John F. Cryan. "Collective Unconscious: How Gut Microbes Shape Human Behavior." *Journal of Psychiatric Research* 63 (2015): 1–9.

Donald, William W. "The Biology of the Canada Thistle (*Cirsium arvense*)." *Review of Weed Science* 6 (1994): 77–101.

Doolittle, W. Ford. 2000. "Uprooting the Tree of Life." *Scientific American* 282, no. 2 (2000): 90–95.

Dornbush, Matthew E., and Philip G. Hahn. "Consumers and Establishment Limitations Contribute More Than Competitive Interactions in Sustaining Dominance of the Exotic Herb Garlic Mustard in a Wisconsin, USA Forest." *Biological Invasions* 15 (2013): 2691–2706.

Dornhaus, Anna, and Nigel R. Franks. "Individual and Collective Cognition in Ants and Other Insects (Hymenoptera: Formicidae)." *Myrmecological News* 11 (2008): 215–226.

Dunn, Rob R. *Never Home Alone: From Microbes to Millipedes, Camel Crickets and Honey Bees, the Natural History of Where We Live.* New York: Basic Books, 2018.

Ebeling, Walter. *Urban Entomology.* Berkeley: Division of Agricultural Sciences, University of California, 1978.

Edgecombe, Gregory D., and Gonzalo Giribet. "Evolutionary Biology of Centipedes (Myriapoda: Chilopoda)." *Annual Review of Entomology* 52 (2007): 151–170.

Emlen, Douglas J. "Alternative Reproductive Tactics and Male-Dimorphism in the Horned Beetle *Onthophagus acuminatus* (Coleoptera: Scarabaeidae)." *Behavioral Ecology and Sociobiology* 41, no. 5 (1997): 335–341.

Erwin, Alexis C., Tobias Züst, Jared G. Ali, and Anurag A. Agrawal. "Above-Ground Herbivory by Red Milkweed Beetles Facilitates Above- and Below-Ground Conspecific Insects and Reduces Fruit Production in Common Milkweed." *Journal of Ecology* 102 (2014): 1038–1047.

Estes, Anne M., David J. Hearn, Emilie C. Snell-Rood, Michele Feindler, Karla Feeser, Tselotie Abebe, Julie C. Dunning Hotopp, and Armin P. Moczek. "Brood Ball-Mediated Transmission of Microbiome Members in the Dung Beetle, *Onthophagus taurus* (Coleoptera: Scarabaeidae)." *PLoS ONE* 8, no. 11 (2013): e79061. Doi:10.1371/journal.pone 0079061.

Fischer, Thilo C. "Caterpillars and Cases of Tineidae (Clothes Moths, Lepidoptera) from Baltic Amber (Eocene)." *Zitteliana A* 54 (2014): 75–81.

Flatt, Thomas, and Wolfgang W. Weisser. "The Effects of Mutualistic Ants on Aphid Life History Traits." *Ecology* 81, no. 12 (2000): 3522–3529.

Food and Agriculture Organization of the United Nations (FAOSTAT). "Pesticides Use"; https://www.fao.org/faostat/en/#data/RP.

Fox, Douglas. "Lakes Under the Ice: Antarctica's Secret Garden." *Nature* 512, no. 7514 (2014): 244–246.

Franks, Nigel R., Stephen C. Pratt, Eamonn B. Mallon, Nicholas F. Britton, and David J. T. Sumpter. "Information Flow, Opinion Polling and Collective Intelligence in House-Hunting Social Insects." *Philosophical Transactions: Biological Sciences* 357, no. 1427 (2002): 1567–1583.

Furness, Amber N., and Daniel A. Soluk. "The Potential of Diversion Structures to Reduce Roadway Mortality of the Endangered Hine's

Emerald Dragonfly (*Somatochlora hineana*)." *Journal of Insect Conservation* 19, no. 3 (2015): 449–455.

Gandhi, Mohandas K. *Hindu Dharma.* Delhi: Orient Paperbacks, 1978.

Gerdes, Antje B. M., Gabriele Uhl, and Georg W. Alpers. "Spiders Are Special: Fear and Disgust Evoked by Pictures of Arthropods." *Evolution and Human Behavior* 30 (2009): 66–73.

Gilbert, Francis. "The Evolution of Imperfect Mimicry in Hoverflies." In *Insect Evolutionary Ecology: Proceedings of the Royal Entomology Society's 22nd Symposium*, edited by Mark D. E. Fellowes, Graham J. Holloway, and Jens Rolff (2005): 231–288.

Gill, Steven R., Mihai Pop, Robert T. DeBoy, Paul B. Eckburg, Peter J. Turnbaugh, Buck S. Samuel, Jeffrey I. Gordon, David A. Relman, Claire M. Fraser-Liggett, and Karen E. Nelson. "Metagenomic Analysis of the Human Distal Gut Microbiome." *Science* 312, no. 5778 (2006): 1355–1359.

Girardin, Barbara W., and Sandra Steveson. "Millipedes—Health Consequences." *Journal of Emergency Nursing* 28 (2002): 107–110.

Gluesenkamp, Daniel, Michael Chassé, Mark Frey, V. Thomas Parker, Michael C. Vasey, and Betty Young. "Back from the Brink: A Second Chance at Discovery and Conservation of the Franciscan Manzanita." *Fremontia* 37, no. 4/38, no. 1 (2009/2010): 3–17.

Global Invasive Species Database. Invasive Species Specialist Group; www.iucngisd.org/gisd/species.php?sc=109.

González, Grizelle, Christina M. Murphy, and Juliana Belén. "Direct and Indirect Effects of Millipedes on the Decay of Litter of Varying Lignin Content." *Tropical Forests* 2 (2012): 37–50.

González-Teuber, Marcia, Martin Kaltenpoth, and Wilhelm Boland. "Mutualistic Ants as an Indirect Defence Against Leaf Pathogens." *New Phytologist* 202, no. 2 (2014): 640–650.

Goodpaster, Kenneth E. "On Being Morally Considerable." *The Journal of Philosophy* 75, no. 6 (1978): 308–325.

Gould, James L., and Carol Grant Gould. *The Honey Bee.* New York: Scientific American Library, 1995.

Goulson, Dave, Gillian C. Lye, and Ben Darvill. "Decline and Conservation of Bumble Bees." *Annual Review of Entomology* 53 (2008): 191–208.

Graystock, Peter, Kathryn Yates, Sophie E. F. Evison, Ben Darvill, Dave Goulson, and William O. H. Hughes. "The Trojan Hives: Pollinator Pathogens, Imported and Distributed in Bumblebee Colonies." *Journal of Applied Ecology* 50 (2013): 1207–1215.

Greenspan, Ralph J., and Jean-Francois Ferveur. "Courtship in *Drosophila*." *Annual Review of Genetics* 34 (2000): 205–232.

Grim, John. "Ecology and Religion: Ecology and Indigenous Traditions." In *Encyclopedia of Religion*, 2nd edition, volume 4, edited by Lindsay Jones, 2616–2620. Detroit: Macmillan Reference USA, 2005.

Grim, John, and Mary Evelyn Tucker. *Ecology and Religion*. Washington, D.C.: Island Press, 2014.

Grimaldi, David, and Michael S. Engel. *Evolution of the Insects*. New York: Cambridge University Press, 2005.

Grotewiel, Michael S., Ian Martin, Poonam Bhandari, and Eric Cook-Wiens. "Functional Senescence in *Drosophila melanogaster*." *Ageing Research Reviews* 4 (2005): 372–397.

Grove, Simon J. "Saproxylic Insect Ecology and the Sustainable Management of Forests." *Annual Review of Ecological Systems* 33 (2002): 1–23.

Guggisberg, Alessia, Erik Welk, René Sforza, David P. Horvath, James V. Anderson, Michael E. Foley, and Loren H. Rieseberg. "Invasion History of North American Canada Thistle, *Cirsium arvense*." *Journal of Biogeography* 39, no. 10 (2012): 1919–1931.

Guillem, Rhian M., Falko P. Drijfhout, and Stephen J. Martin. "Using Chemo-Taxonomy of Host Ants to Help Conserve the Large Blue Butterfly." *Biological Conservation* 148, no. 1 (2012): 39–43.

Gundry, Jamie, and Charles Ellington. "The Kinematics of Hoverfly Flight." *Comparative Biochemistry and Physiology* Abstract A6.26, 146 (2007): S115.

Gunter, Nicole L., Tom A. Weir, Adam Slipinksi, Ladislav Bocak, and Stephen L. Cameron. "If Dung Beetles (Scarabaeidae: Scarabaeinae) Arose in Association with Dinosaurs, Did They Also Suffer a Mass Co-Extinction at the K-Pg Boundary?" *PLoS ONE* 11, no. 5 (2016): e0153570. Doi:10.1371/journal.pone.0153570.

Hallman, Caspar A., Martin Sorg, Eelke Jongejans, Henk Siepel, Nick Hofland, Heinz Schwan, Werner Stenmans, Andreas Müller, Hubert Sumser, Thomas Hörren, Dave Goulson, and Hans de Kroon. "More Than 75 Percent Decline over 27 Years in Total Flying Insect Biomass in Protected Areas." PLoS ONE 12(10): e0185809; https://doi.org/10.1371/journal.pone.0185809.

Haq, S. Nomanul. "Islam and Ecology: Toward Retrieval and Reconstruction." *Daedalus* 130, no. 4 (2001): 141–177.

Hill, Clive. "Introducing the Tree Bumblebee *Bombus hypnorum*." *Bee Craft* 95, no. 5 (2013): 17–19.

Hodgson, Jesse M. *The Nature, Ecology, and Control of Canada Thistle*. United States Department of Agriculture Technical Bulletin 1386. Washington, D.C.: United States Government Printing Office, 1968.

Hoehl, Stefanie, Kahl Hellmer, Maria Johansson, and Gustaf Gredebäck. "Itsy Bitsy Spider . . . : Infants React with Increased Arousal to Spiders and Snakes." *Frontiers in Psychology* 8 (2017): 1710. Doi: 10:3389/fpsyg .2017.01710.

Hogan, Michael C. "*Arctostaphylos*." *The Encyclopedia of Earth*. www.eoearth .org/view/article/150218, accessed July 8, 2017.

Hogg, Brian N., Erik H. Nelson, Nicholas J. Mills, and Kent M. Daane. "Floral Resources Enhance Aphid Suppression by a Hoverfly." *Entomologia Experimentalis et Applicata* 141 (2011): 138–144.

Hölldobler, Bert, and Edward O. Wilson. *The Superorganism: The Beauty, Elegance, and Strangeness of Insect Societies*. New York: W.W. Norton. 2009.

Holloway, Alisha K., and Gary D. Schnell. "Relationship Between Numbers of the Endangered American Burying Beetle *Nicrophorus americanus* Olivier (Coleoptera: Silphidae) and Available Food Resources." *Biological Conservation* 81, nos. 1–2 (1997): 145–152.

Hunt, John, and Leigh W. Simmons. "Behavioural Dynamics of Biparental Care in the Dung Beetle *Onthophagus taurus*." *Animal Behaviour* 64 (2002): 65–75.

Inoue, Maki N., Fuminori Ito, and Koichi Goka. "Queen Execution Increases Relatedness Among Workers of the Invasive Argentine Ant, *Linephithema humile*." *Ecology and Evolution* 5, no. 18 (2015): 4098–4107.

Jauker, Frank, and Volkmar Wolters. "Hover Flies Are Efficient Pollinators of Oilseed Rape." *Oecologia* 156 (2008): 819–823.

Johnson-Davies, Denys. *The Island of Animals*. Austin: University of Texas Press, 1994.

Kellert, Stephen R. *The Value of Life: Biological Diversity and Human Society*. Washington, D.C.: Island Press, 1996.

Kitaoka, T. K., and J. C. Nieh. "Manuscript in Preparation for Behavioral Ecology and Sociobiology Bumble Bee Pollen Foraging Regulation: Role of Pollen Quality, Storage Levels, and Odor." *Behavioral Ecology and Sociobiology* 63, no. 4 (2009): 501–510.

Klimov, Pavel B., and Barry O'Connor. "Is Permanent Parasitism Reversible?— Critical Evidence from Early Evolution of House Dust Mites." *Systematic Biology* 62, no. 3 (2013): 411–423.

Knell, Robert J., and Simmons, Leigh W. "Mating Tactics Determine Patterns of Condition Dependence in a Dimorphic Horned Beetle." *Proceedings of the Royal Society B* 277 (2010): 2347–2353.

Koch, R. L. "The Multicolored Asian Lady Beetle, *Harmonia axyridis:* A Review of Its Biology, Uses in Biological Control, and Non-Target Impacts." *Journal of Insect Science* 3, no. 1 (2003): 32. Doi: 10.1093/jis /3.1.32.

Kozol, Andrea J., Michelle P. Scott, and James F. A. Traniello. "The American Burying Beetle, *Nicrophorus americanus:* Studies on the Natural History of a Declining Species." *Psyche* 95 (1988): 167–176.

Krulwich, Robert. "Three Nice Things We Can Say About Mosquitoes." *Krulwich Wonders: Krulwich on Science* (2008). https://www.npr.org /templates/story/story.php?storyId=93049810, accessed July 8, 2018.

Lankau, Richard A., Victoria Nuzzo, Greg Spyreas, and Adam S. Davis. "Evolutionary Limits Ameliorate the Negative Impact of an Invasive Plant." *Proceedings of the National Academy of Sciences* 106, no. 36 (2009): 15362–15367.

Leong, Misha, Matthew A. Bertone, Amy M. Savage, Keith M. Bayless, Robert R. Dunn, and Michelle D. Trautwein. "The Habitats Humans Provide: Factors Affecting the Diversity and Composition of Arthropods in Houses." *Scientific Reports* 7 (2017): 15347. Doi: 10.1038/s41598-017 -15584-2.

Leopold, Aldo. *A Sand County Almanac.* New York: Oxford University Press, 1949.

Lister, Bradford C., and Andres Garcia. "Climate-Driven Declines in Arthropod Abundance Restructure a Rainforest Food Web." *Proceedings of the National Academy of Sciences* 155, no. 44 (October 2018), E10397– E10406; https://doi.org/10.1073/pnas.1722477115.

LoBue, Vanessa, Megan Bloom Pickard, Kathleen Sherman, Chrystal Axford, and Judy S. DeLoache. "Young Children's Interest in Live Animals." *British Journal of Developmental Psychology* 31 (2012): 57–69.

Lockwood, Jeffrey A. *The Infested Mind: Why Humans Fear, Loathe, and Love Insects.* Oxford: Oxford University Press, 2013.

Lomolino, Mark V., and J. Curtis Creighton. "Habitat Selection, Breeding Success and Conservation of the Endangered American Burying Beetle, *Nicrophorus americanus.*" *Biological Conservation* 77 (1996): 235–241.

Losey, John E., Leslie L. Allee, Erin Stephens, Rebecca R. Smyth, Peter Priolo, Leah Tyrrell, Scott Chaskey, and Leonard Stellwag. "Lady Beetles in New York: Insidious Invasions, Erstwhile Extirpations, and Recent Rediscoveries." *Northeastern Naturalist* 21, no. 2 (2014): 271–284.

Lovejoy, Arthur O. *The Great Chain of Being: A Study of the History of an Idea.* New York: Harper & Row, 1960.

Lowenstein, David M., Kevin C. Matteson, Iyan Xiao, Alexandra M. Silva, and Emily S. Minor. "Humans, Bees, and Pollination Services in the City: The Case of Chicago, IL (USA)." *Biodiversity and Conservation* 23, no. 11 (2014): 2857–2874.

Maimonides. *The Eight Chapters of Maimonides on Ethics,* translated by Joseph I. Gorfinkle. New York: Columbia University Press, 1912.

Malcolm, Stephen B. "Cardenolide-Mediated Interactions Between Plants and Herbivores." In *Herbivores: Their Interaction with Secondary Plant Metabolites,* 2nd edition, volume 1, *The Chemical Participants,* edited by Gerald A. Rosenthal and May Berenbaum, 251–296. San Diego: Academic Press, 1991.

Mallis, Arnold. *Handbook of Pest Control: The Behavior, Life History, and Control of Household Pests,* 6th edition. Cleveland: Franzak and Foster Company, 1982.

Mansourian, Suzan, Anders Enjin, Erling V. Jirle, Vedika Ramesh, Guillermo Rehermann, Paul G. Becher, John E. Pool, and Marcus C. Stensmyr. "Wild African *Drosophila melanogaster* Are Seasonal Specialists on Marula Fruit." *Current Biology* 28 (2018): 3960–3968.

Manton, S. M. "The Evolution of Arthopodan Locomotory Mechanism. Part 8. Functional Requirements and Body Design in Chilopoda, Together with a Comparative Account of Their Skeleton-Muscular Systems and an Appendix on a Comparison Between Burrowing Forces of Annelids and Chilopods and Its Bearing upon the Evolution of the Arthopodan Haemocoel." *Journal of the Linnean Society (Zoology)* 46, nos. 306–307 (1965): 251–497.

Markos, Satci, Lena C. Hileman, Michael C. Vasey, and V. Thomas Parker. "Phylogeny of the *Arctostaphylos hookeri* Complex (Ericaceae) Based on nrDNA Data." *Madrono* 45, no. 3 (1998): 187–199.

Markow, Therese Ann, and Patrick M. O'Grady. "Drosophila Biology in the Genomic Age." *Genetics* 177 (2007): 1269–1276.

McKenna, Duane D., Katherine M. McKenna, Stephen B. Malcom, and May R. Berenbaum. "Mortality of Lepidoptera Along Roadways in Central Illinois." *Journal of the Lepidopterists' Society* 55, no. 2 (2001): 63–68.

Mitich, Larry W. "Crabgrass." *Weed Technology* 2, no. 1 (1988): 114–115.

Mitich, Larry W. "Common Dandelion: The Lion's Tooth." *Weed Technology* 3, no. 3 (1989): 537–539.

Moczek, Armin P., and Jeffrey Cochrane. "Intraspecific Female Brood Parasitism in the Dung Beetle *Onthophagus taurus.*" *Ecological Entomology* 31 (2006): 1–6.

Mondor, E. B., Jay A. Rosenheim, and J. F. Addicott. "Mutualist-Induced Transgenerational Polyphenisms in Cotton Aphid Populations." *Functional Ecology* 22 (2008): 157–162.

Monroe, Emy M., and Hugh B. Britten. "Conservation in Hine's Sight: The Conservation Genetics of the Federally Endangered Hine's Emerald Dragonfly, *Somatochlora hineana.*" *Journal of Insect Conservation* 18, no. 3 (2014): 353–363.

Nash, David R., Thomas D. Als, Roland Maile, Graeme R. Jones, and Jacobus J. Boomsma. "A Mosaic of Chemical Coevolution in a Large Blue Butterfly." *Science* 319, no. 5859 (2008): 88–90.

Nasr, Seyyed H. "Islam, the Contemporary Islamic World and the Environmental Crisis." In *Islam and Ecology: A Bestowed Trust*, edited by Richard C. Foltz, Frederick M. Denny, and Azizan Baharuddin, 85–106. Cambridge: Harvard University Press, 2003.

New, Joshua J., and Tamsin C. German, "Spiders at the Cocktail Party: An Ancestral Threat that Surmounts Inattentional Blindness." *Evolution and Human Behavior* 36, no. 3 (2015): 165–173.

Nhat Hanh, T. "The Sun My Heart." In *Engaged Buddhist Reader*, edited by Arnold Kotler, 162–170. Berkeley, Calif.: Parallax Press, 1996.

Nhat Hanh, T. *No Death, No Fear: Comforting Wisdom for Life.* New York: Riverhead Books, 2002.

Nhat Hanh, T. *Love Letter to the Earth.* Berkeley, Calif.: Parallax Press, 2013.

NIH Human Microbiome Project; https://hmpdacc.org/hmp/overview, accessed July 5, 2018.

North American Pollinator Protection Campaign Bombus Task Force. "Bumblebees Are Essential." http://www.pollinator.org/PDFs/NAPPC.BumbleBee.brochFINAL.pdf, accessed July 10, 2017.

Nowakowski, Matilda E., Randi McCabe, Karen Rowa, Joe Pellizzari, Michael Surette, Paul Moayyedi, and Rebecca Anglin. "The Gut Microbiome: Potential Innovations for the Understanding and Treatment of Psychopathology." *Canadian Psychology* 57, no. 2 (2016): 67–75.

Nuzzo, Victoria. "Element Stewardship Abstract for *Alliaria petiolata* (*Alliaria officinalis*) Garlic Mustard." Arlington, Va., Nature Conservancy, 2002; https://www.invasive.org/weedcd/pdfs/tncweeds/allipet.pdf.

Nuzzo, Victoria A. "Experimental Control of Garlic Mustard (*Alliaria petiolata* [Bieb.] Cavara and Grande) in Northern Illinois Using Fire, Herbicide, and Cutting." *Natural Areas Journal* 11 (1991): 158–167.

Nuzzo, Victoria A., John C. Maerz, and Bernd Blossey. "Earthworm Invasion as the Driving Force Behind Plant Invasion and Community Change in Northeastern North American Forests." *Conservation Biology* 23 (2009): 966–974.

Oliver, Randy. "2012 Almond Pollination Update." ScientificBeekeeping.com: Beekeeping Through the Eyes of a Biologist, 2012. http://scientificbeekeeping.com/2012-almond-pollination-update.

Online Etymology Dictionary. Bug. https://www.etymonline.com/word/bug, accessed July 4, 2018.

Origen. *On First Principles: Being Koetschau's Text of the De Principiis.* Translated by George W. Butterworth. New York: Harper and Row, 1966.

Ostfeld, Richard S., Charles D. Canham, and Stephen R. Pugh. "Intrinsic Density-Dependent Regulation of Vole Populations." *Nature* 366, no. 18 (1993): 259–261.

Palmer, Todd M., Maureen L. Stanton, Truman P. Young, Jacob R. Goheen, Robert M. Pringle, and Richard Karban. "Breakdown of an Ant-Plant Mutualism Follows the Loss of Large Herbivores from an African Savanna." *Science* 319, no. 5860 (2008): 192–195.

Penney, Heather D., Christopher Hassall, Jeffrey H. Skevington, Brent Lamborn, and Thomas N. Sherratt. "The Relationship Between Morphological and Behavioral Mimicry in Hover Flies (Diptera: Syrphidae)." *The American Naturalist* 183, no. 2 (2014): 281–289.

Pennisi, Elizabeth. "Africa's Soil Engineers: Termites." *Science* 347, no. 6222 (2015): 596–597.

Pintor, Lauren M., and Daniel A. Soluk. "Evaluating the Non-Consumptive, Positive Effects of a Predator in the Persistence of an Endangered Species." *Biological Conservation* 130 (2006): 584–591.

Plarre, Rudy, and Bianca Krüger-Carstensen. "An Attempt to Reconstruct the Natural and Cultural History of the Webbing Clothes Moth *Tineola bisselliella* Hummel (Lepidoptera: Tineidae)." *Journal of Entomological and Acarological Research Series II* 43, no. 2 (2011): 83–93.

Powell, Bradford E., and Jules Silverman. "Population Growth of *Aphis gossypii* and *Myzus persicae* (Hemipters: Aphididae) in the Presence of

Linepithema humile and *Tapinoma sessile* (Hymenoptera: Formicidae)." *Environmental Entomology* 39, no. 5 (2010): 1492–1499.

Preheim, L. Misha. "Biophilia, the Endangered Species Act, and a New Endangered Species Paradigm." *William and Mary Law Review* 42, no. 3 (2001): 1053–1076.

Pringle, Robert M., Daniel F. Doak, Alison K. Brody, Rudy Jocqué, and Todd M. Palmer. "Spatial Pattern Enhances Ecosystem Functioning in an African Savanna." *PLoS Biology* 8, no. 5 (2010): e1000377; https://doi.org/10.1371/journal.pbio.1000377.

Punzo, Fred. "Analysis of Maze Learning in the Silverfish, *Lepisma saccharina* (Thysanura: Lepismatidae)." *Journal of the Kansas Entomological Society* 53, no. 3 (1980): 653–661.

Quammen, David. *The Flight of the Iguana: A Sidelong View of Science and Nature.* New York: Delacorte Press, 1988.

Rasmann, Sergio, M. Daisy Johnson, and Anurag A. Agrawal. 2009. "Induced Responses to Herbivory and Jasmonate in Three Milkweed Species." *Journal of Chemical Ecology* 35, no. 11 (2009): 1326–1334.

Ratcliffe, B. C. "Endangered American Burying Beetle Update." 1997; www.museum.unl.edu/research/entomology/endanger.htm.

Raven, Peter H., Ray F. Evert, and Susan F. Eichhorn. *The Biology of Plants,* 7th edition. New York: W.H. Freeman, 2005.

Reemer, Menno. "Saproxlyic Hoverflies Benefit by Modern Forest Management (Diptera: Syrphidae)." *Journal of Insect Conservation* 9 (2005): 49–59.

Reinhardt, Klaus, and Michael T. Siva-Jothy. "Biology of the Bed Bugs (Cimicidae)." *Annual Review of Entomology* 52 (2007): 351–374.

Reuters. "On This Day: Obituary, Albert Schweitzer, 90, Dies at His Hospital"; movies2.nytimes.com/learning/general/onthisday/bday/0114.html.

Robinson, William H. *Urban Insects and Arachnids.* Cambridge: Cambridge University Press, 2005.

Rodgers, Vikki L., Kristina A. Stinson, and Adrien C. Finzi. "Ready or Not, Garlic Mustard Is Moving In: *Alliaria petiolata* as a Member of Eastern North American Forests." *BioScience* 58, no. 5 (2008): 426–436.

Rolston, Holmes, III. "Naturalizing Values: Organisms and Species." In *Environmental Ethics: Readings in Theory and Application,* edited by Louis P. Pojman, 76–89. Belmont, Calif.: Wadsworth Publishing/Thomson Learning, 2001.

Rolston, Holmes, III (1996). "Feeding People Versus Saving Nature?" In

World Hunger and Morality, edited by William Aiken and Hugh LaFollette, 248–267. Englewood Cliffs, N.J.: Prentice-Hall, 1996.

Rooney, Thomas P., and David A. Rogers. "Colonization and Effects of Garlic Mustard (*Alliaria petiolata*), European Buckthorn (*Rhamnus cathartica*), and Bell's Honeysuckle (*Lonicera x bella*) on Understory Plants After Five Decades in Southern Wisconsin Forests." *Invasive Plant Science and Management* 4, no. 3 (2011): 317–325.

Roques, Alain. "Species Factsheet: *Aphis gossypii.*" *Delivering Alien Invasive Species Inventories for Europe.* 2006; http://www.europe-aliens.org/species Factsheet.do?speciesId=50793.

Rose, Michael, and Brian Charlesworth. "A Test of Evolutionary Theories of Senescence." *Nature* 287 (1980): 141–142.

Rotheray, Graham, and Francis Gilbert. "Phylogeny of Palearctic Syrphidae (Diptera) Evidence from Larval Stages." *Zoological Journal of the Linnean Society* 127 (1999): 1–112.

Rozin, Paul, and April E. Fallon. "A Perspective on Disgust." *Psychological Review* 94, no. 1 (1987): 23–41.

Salvucci, Emiliano. "Selfishness, Warfare and Economics, or Integration, Cooperation and Biology." *Frontiers in Cellular and Infection Microbiology* 2 (2012): 54; http://doi.org/10.3389/fcimb.2012.00054.

Sanchez, Anita. *The Teeth of the Lion: The Story of the Beloved and Despised Dandelion.* Blacksburg, Va.: McDonald and Woodward, 2006.

Sánchez-Bayo, Francisco, and Kris A. G. Wyckhuys. "Worldwide Decline of the Entomofauna: A Review of Its Drivers." *Biological Conservation* 232 (2019): 8–27.

Schmidt, Marjorie G. "Natives for Your Garden: San Francisco Manzanita *Arctostaphylos hookeri* subsp. *Franciscana.*" *Fremontia* 5, no. 4 (1978): 38–39.

Schwarz, Horst H., Martin Starrach, and Stella Koulianos. "Host Specificity and Permanence of Associations Between Mesostimatic Mites (Acari: Anactinotrichida) and Burying Beetles (Coleoptera: Silphidae: *Nicrophorus*)." *Journal of Natural History* 32, no. 2 (1998): 159–172.

Schweitzer, A. (1946). *Civilization and Ethics.* London: Adam and Charles Black.

Scott, Michelle Pellissier. "The Ecology and Behavior of Burying Beetles." *Annual Review of Entomology* 43 (1998): 595–618.

Shear, William A. "The Chemical Defenses of Millipedes (Diplopoda): Biochemistry, Physiology and Ecology." *Biochemical Systematics and Ecology* 61 (2015): 78–117.

Shik, Jonathan Z., Adam D. Kay, and Jules Silverman. "Aphid Honeydew Provides a Nutritionally Balanced Resource for Incipient Argentine Ant Mutualists." *Animal Behaviour* 95 (2014): 33–39.

Sierwald, Petra, and Jason E. Bond. "Current Status of the Myriapod Class Diplopoda (Millipedes): Taxonomic Diversity and Phylogeny." *Annual Review of Entomology* 52 (2007): 401–420.

Silverman, Jules, and Hannaliese Selbach, "Feeding Behavior and Survival of Glucose-Averse *Blatella germanica* (Orthoptera: Blatoidea: Blattellidae) Provided Glucose as a Sole Food Source." *Journal of Insect Behavior* 11, no. 1 (1998): 93–102.

Singer, Peter. *Animal Liberation.* London: Jonathan Cape, 1990.

Singer, Peter. "Are Insects Conscious?" *Project Syndicate*, May 12, 2016; https://www.project-syndicate.org/commentary/are-insects-conscious-by-peter-singer-2016-05.

Six Nations. "A Basic Call to Consciousness: The Haudenosaunee Address to the Western World." In *Basic Call to Consciousness*, edited by Akwasane Notes. Summertown, Tenn.: Book Publishing Company, 1978.

Slade, Eleanor M., Terhi Riutta, Tomas Roslin, and Hanna L. Tuomisto. "The Role of Dung Beetles in Reducing Greenhouse Gas Emissions from Cattle Farming." *Scientific Reports* 6 (2016): 18140. Doi: 10.1038/srep18140.

Smigel, Jacob T., and Allen G. Gibbs. "Conglobation in the Pill Bug, *Armadillidium vulgare*, as a Water Conservation Mechanism." *Journal of Insect Science* 8, no. 44 (2008): 1–9.

Smith, David Joseph, and Dale Warren Griffin. "Inadequate Methods and Questionable Conclusions in Atmospheric Life Study." *Proceedings of the National Academy of Sciences* 110, no. 23 (2013): E2084; http://doi.org/10.1073/pnas.1302612110.

Smith, Dena M., and Jonathan D. Marcot. "The Fossil Record and Macro-evolutionary History of the Beetles." *Proceedings of the Royal Society B* 282 (2015): 20150060; http://dxdoi.org/10.1098/rspb.2015.0060.

Solbrig, Otto T. "The Population Biology of Dandelions: These Common Weeds Provide Experimental Evidence for a New Model to Explain the Distribution of Plants." *American Scientist* 59, no. 6 (1971): 686–694.

Stadler, Bernhard, and Anthony F. G. Dixon. "Ecology and Evolution of Aphid-Ant Interactions." *Annual Review of Ecology, Evolution, and Systematics* 36 (2005): 345–372.

Stubbs, C. S. 2002. "Commercial Bumble Bee (*Bombus impatiens*) Management for Wild Blueberry Pollination." University of Maine Extension

Fact Sheet Number 302, 2002; https://extension.umaine.edu/blueberries/factsheets/bees/302-commercial-bumble-bee-bombus-impatiens-management-for-wild-blueberry-pollination.

Taylor, Paul. "The Ethics of Respect for Nature." *Environmental Ethics* 3, no. 3 (1981): 197–218.

Theis, Nina. "Fragrance of Canada Thistle (*Cirsium arvense*) Attracts Both Floral Herbivores and Pollinators." *Journal of Chemical Ecology* 32 (2006): 917–927.

Tiley, Gordon E. D. "Biological Flora of the British Isles: *Cirsium arvense* (L.) Scop." *Journal of Ecology* 98 (2010): 938–983.

Tirosh-Samuelson, Hava. "Nature in the Sources of Judaism." *Daedalus* 130 no. 4 (2001): 99–124.

Tsutsui, Neil D., Andrew V. Suarez, David A. Holway, and Ted J. Case. "Reduced Genetic Variation and the Success of an Invasive Species." *Proceedings of the National Academy of Sciences* 97, no. 11 (2000): 5948–5953.

Undheim, Eivind A. B., Gryan G. Fry, and Glenn F. King. "Centipede Venom: Recent Discoveries and Current State of Knowledge." *Toxins* 7 (2015): 679–704.

Urbanski, Jennifer, Motoyoshi Mogi, Deborah O'Donnell, Mark DeCotilis, Takako Toma, and Peter Armbruster. "Rapid Adaptive Evolution of Photoperiodic Response During Invasion and Range Expansion Across a Climatic Gradient." *The American Naturalist* 179, no. 4 (2012): 490–500.

U.S. Congress. *Endangered Species Act of 1973*. Washington, D.C.: U.S. Government Publishing Office, 1973; www.gpo.gov/fdsys/pkg/STATUTE-87/pdf/STATUTE-87-Pg884.pdf; accessed July 12, 2018.

U.S. EPA, Pesticides Industry Sales and Usage: 2008–2012 Market Estimates, Tables 3.2, 4.4; https://www.epa.gov/sites/production/files/2017-01/documents/pesticides-industry-sales-usage-2016_0.pdf.

U.S. Fish and Wildlife Service. "Endangered and Threatened Wildlife and Plants: Designation of Critical Habitat for *Arctostaphylos franciscana* (Franciscan Manzanita)," 2013; https://www.federalregister.gov/documents/2013/12/20/2013-30165/endangered-and-threatened-wildlife-and-plants-designation-of-critical-habitat-for-arctostaphylos.

U.S. Fish and Wildlife Service. "American Burying Beetle Impact Assessment for Project Review," 2014; https://www.fws.gov/southwest/es/oklahoma/documents/abb/abb%20impact%20assessment%20for%20project%20reviews_6mar2014.pdf.

U.S. Fish and Wildlife Service. "Reclassifying the American Burying Beetle from Endangered to Threatened on the Federal List of Endangered and Threatened Wildlife with a 4(d) Rule," 2019. https://s3.amazonaws.com/public-inspection.federalregister.gov/2019-09035.pdf.

U.S. Fish and Wildlife Service. "Hine's Emerald Dragonfly (*Somatochlora hineana*) Recovery Plan," 2001. https://www.fws.gov/midwest/endangered/insects/hed/pdf/hedplan.pdf.

U.S. Fish and Wildlife Service. Recovery Outline for the *Artcostaphylos franciscana* (Franciscan Manzanita). Sacramento, California, 2013. https://esadocs.ccidev.org/ESAdocs/recovery_plan/Franciscan_manzanita_Recovery_Outline.pdf.

van Wilgenburg, Ellen, Candice W. Torres, and Neil D. Tsutsui. "The Global Expansion of a Single Ant Supercolony." *Evolutionary Applications* 3, no. 2 (2010): 136–143.

Vetter, Richard S. "Arachnophobic Entomologists: When Two More Legs Makes a Big Difference." *American Entomologist* 59, no. 3 (2013): 168–175.

Walter, Jens. "Ecological Role of Lactobacilli in the Gastrointestinal Tract: Implications for Fundamental and Biomedical Research." *Applied and Environmental Microbiology* 74, no. 16 (2008): 4985–4996.

Warrant, Eric, and Marie Dacke. "Visual Orientation and Navigation in Nocturnal Arthropods." *Brain, Behavior and Evolution* 75 (2010): 156–173.

Weems, Howard V. "Featured Creatures: A Hover Fly." University of Florida Publication No. EENY-185 (2014). http://entnemdept.ufl.edu/creatures/beneficial/hover_fly.htm.

Weghofer, Margit, Monika Grote, Yvonne Resch, Anne Casset, Michael Kneidinger, Iolanta Kopec, Wayne R. Thomas, Enrique Fernández-Caldas, Michael Kabesch, Rosetta Ferrara, Adriano Mari, Ashok Purohit, Gabrielle Pauli, Friedrich Horak, Walter Keller, Peter Valent, Rudolf Valenta, and Susanne Vrtala. "Identification of Der p 23, a Peritrophin-Like Protein, as a New Major *Dermatophagoides pteronyssinus* Allergen Associated with the Peritrophic Matrix of Mite Fecal Pellets." *Journal of Immunology* 190 (2013): 3059–3067.

Wells, H. G. *The Empire of the Ants and Other Stories.* New York: Scholastic Book Services, 1977.

Wexler, Hannah M. "*Bacteroides:* The Good, the Bad, and the Nitty-Gritty." *Clinical Microbiology Reviews* 20, no. 4 (2007): 593–621.

White, Anthony J., Stephen D. Wratten, Nadine A. Berry, and Ursula Weigmann. "Habitat Manipulation to Enhance Biological Control of Brassica Pests by Hover Flies (Diptera: Syrphidae). *Journal of Economic Entomology* 88, no. 5 (1995): 1171–1176.

White, Lynn, Jr. "The Historical Roots of Our Ecologic Crisis." *Science* 155, no. 3767 (1967): 1203–1207.

"Wild Equity Institute Versus Ken Salazar, Secretary of the Interior and the United States Fish and Wildlife Service." United States District Court, Northern District of California San Francisco Division. June 14, 2011. http://www.trialinsider.com/pdf/manzanitaESA.pdf.

Wilson, Edward O. *Biophilia.* Boston: Harvard University Press, 2009.

Wilson, Edward O., and Bert Hölldobler. "The Rise of the Ants: A Phylogenetic and Ecological Explanation." *Proceedings of the National Academy of Sciences* 102, no. 21 (2005): 7411–7414.

Wootton, Robin J. "Functional Morphology of Insect Wings." *Annual Review of Entomology* 37 (1992): 113–140.

Worobey, John, Dina M. Fonseca, Carolina Espinosa, Sean Healy, and Randy Gaugler. "Child Outdoor Physical Activity Is Reduced by Prevalence of the Asian Tiger Mosquito, Aedes albopictus." *Journal of the American Mosquito Control Association* 29, no. 1 (2013): 78–80.

Wright, Brianna, and Daniel B. Tinker. "Canada Thistle (*Cirsium arvense* (L) Scop.) Dynamics in Young, Postfire Forests in Yellowstone National Park, Northwestern Wyoming." *Plant Ecology* 213, no. 4 (2012): 613–624.

Young, Orrey P. "The Biology of the Silphidae: A Coded Bibliography." *Maryland Agricultural Experiment Station, Miscellaneous Publication* 981 (1983): 1–48.

Zalasiewicz, Jan, Mark Williams, Will Steffen, and Paul Crutzen. "The New World of the Anthropocene." *Environmental Science and Technology* 44, no. 7 (2010): 2228–2231.

Zalucki, Myron P., Stephen B. Malcolm, Timothy D. Paine, Christopher C. Hanlon, Lincoln P. Brower, and Anthony R. Clarke. "It's the First Bites that Count: Survival of First-Instar Monarchs on Milkweeds." *Austral Ecology* 26 (2001): 1–9.

Zhou, Bin, Chui-Hua Kong, Yong-Hua Li, Peng Wang, and Xiao-Hua Xu. "Crabgrass (*Digitaria sanguinalis*) Allelochemicals that Interfere with Crop Growth and the Soil Microbial Community." *Journal of Agricultural and Food Chemistry* 61, no. 22 (2013): 5310–5317.

Ziska, Lewis H. "The Impact of Nitrogen Supply on the Potential Response of a Noxious, Invasive Weed, Canada Thistle (*Cirsium arvense*) to Recent Increases in Atmospheric Carbon Dioxide." *Physiologia Plantarum* 119 (2003): 105–112.

Zwarts, Liesbeth, Marijke Versteven, and Patrick Callaerts. "Genetics and Neurobiology of Aggression in *Drosophila*." *Fly* 6, no. 1 (2012): 35–48.

Index